Statistical Analysis of Spatial and Spatio-Temporal Point Patterns

Third Edition

MONOGRAPHS ON STATISTICS AND APPLIED PROBABILITY

General Editors

F. Bunea, V. Isham, N. Keiding, T. Louis, R. L. Smith, and H. Tong

Monographs on Statistics and Applied Probability 128

Statistical Analysis of Spatial and Spatio-Temporal Point Patterns
Third Edition

Peter J. Diggle
Lancaster University
England, UK

CRC Press
Taylor & Francis Group
Boca Raton London New York

CRC Press is an imprint of the
Taylor & Francis Group, an **informa** business

A CHAPMAN & HALL BOOK

CRC Press
Taylor & Francis Group
6000 Broken Sound Parkway NW, Suite 300
Boca Raton, FL 33487-2742

First issued in paperback 2022

ISBN 13: 978-1-03-247747-3 (pbk)
ISBN 13: 978-1-4665-6023-9 (hbk)

DOI: 10.1201/b15326

Visit the Taylor & Francis Web site at
http://www.taylorandfrancis.com

and the CRC Press Web site at
http://www.crcpress.com

To the memory of Julian Besag FRS, 1945-2010

Contents

List of Figures

List of Tables

Preface

A spatial point pattern is a set of locations, irregularly distributed within a designated region and presumed to have been generated by some form of stochastic mechanism. In most applications, the designated region is essentially planar (two-dimensional Euclidean space), but one-dimensional applications are also possible, and three-dimensional applications are becoming more common in conjunction with the development of more sophisticated three-dimensional scanning microscopes. The first edition of *Statistical Analysis of Spatial Point Patterns* appeared in 1983. Its aim was to cover the major methodological themes of the subject and their application to data arising in the biological sciences, especially ecology.

In the second edition, published in 2003, I extended the methodological discussion to cover major developments in the intervening years, but also tried to preserve the applied flavour of the book. Much of the newer work in the area tends to be mathematically sophisticated. My aim in covering the newer methodological developments was to discuss the central ideas without going into the full technical detail which a rigorous treatment would require, and which is available in the original articles. I also resisted the temptation to discuss spatial statistics more widely. Cressie (1991) identified the three main branches of spatial statistics as geostatistics (spatially continuous processes), lattice processes (spatially discrete processes) and spatial point processes. Whilst these three topics are to some extent inter-linked, they nevertheless give rise to distinct stochastic models and associated statistical methods, and can therefore be studied separately.

Within the realm of spatial point processes, perhaps the most important theoretical development over the last three decades has been the provision of formal, likelihood-based methods of inference for a reasonably wide range of models. These have partially replaced the more *ad hoc* methods which prevailed in the early 1980's. Nevertheless, some of the *ad hoc* methods remain useful, and have themselves been extended in various ways, for example in the adaptation of non-parametric smoothing methods to spatial point processes. New applications have also emerged and, as is usual in statistics, have in turn motivated further methodological development. The two new areas of application on which I draw most heavily, in my own research and in the book, are micro anatomy and epidemiology.

In microanatomy, the points in an observed pattern typically are reference locations for cells in a microscopic tissue section. The underlying cellular structure influences the kinds of models which are appropriate, usually involving the concept of interactions between near-neighbouring cells. Perhaps more fundamentally from a statistical viewpoint, most micro-anatomical studies use a replicated sampling design in which data are obtained from several subjects and/or several tissue sections per subject, in contrast to the traditional emphasis throughout spatial statistics on the analysis of unreplicated patterns.

In epidemiology, the points are reference locations (typically place of residence) for cases of a disease in a geographical region, often supplemented by the reference locations for a set of controls sampled from the underlying population at risk. The principal methodological challenge in this area of application is to use the case-control paradigm to circumvent the obvious difficulty of building credible parametric models for human population distributions in a heterogeneous environment.

When the first edition was written, there were few other books available on the then-emerging subject of spatial statistics, and none at all which dealt exclusively with the statistical analysis of spatial point patterns. This is no longer the case. Books on spatial point processes and associated statistical methods include Møller and Waagepetersen (2004) and Ilian *et al* (2008), whilst Van Lieshout (2000) deals exclusively with Markov point processes and their statistical analysis. The first general text on spatial statistics was Ripley (1981). This, and Cressie (1991), remain standard references. Other books on spatial statistics that include substantial material on point processes include the two volumes of Upton and Fingleton (1985, 1989), Cliff and Ord (1981), Bailey and Gatrell (1995), Stoyan, Kendall and Mecke (1995), Waller and Gotway (2004), Schabenberger and Gotway (2004), and Gelfand *et al.* (2010). Additionally, and typically for a maturing field, a wide variety of more specialised books have also become available. For example, Matérn (1986) is a re-issue of Bertil Matérn's classic 1960 Swedish doctoral dissertation that laid many of the foundations for later developments in spatial point processes and geostatistics. Rue and Held (2005) discuss Gaussian Markov random fields, the most widely used modelling framework for spatially discrete processes. Chilès and Delfiner (1999, 2012) and Diggle and Ribeiro (2007) cover the so-called "classical" and "model-based" approaches to geostatistics, respectively.

Some topics which seemed important at the beginning of the 1980's had become less so by 2003, and I reduced the space given to them. One example was the discussion of methods for sparse sampling of spatial point processes *in situ*. These methods arose in the 1950's and 1960's, principally in connection with field-work by ecologists, who needed to investigate the density and spatial pattern of plant communities in the field. By 2003, these methods were rarely used because technological advances had made much more straighforward the task of mappping a spatial point pattern for later analysis using the more sophisticated methods that were by then available. To my surprise, however, these methods subsequently experienced a minor revival when they began to be used for analysing the size and pattern of refugee encampments (see, for example, Bostoen, Chalabi and Grais, 2007).

Thus forewarned, in this new book I have not deleted any material from the second edition, other than to correct a number of errors, but I have added substantial new material in places.

The biggest change from the second edition, reflected in the enlarged title, is to discuss spatio-temporal point patterns. Spatio-temporal point process data have long been studied in specialised fields, notably seismology (see, for example, Zhuang, Ogata and Vere-Jones, 2002). However, in the last decade there has been an acceleration of methodological development, accompanied by a diversification of application as spatio-temporally indexed data have become more widely available in many scientific fields. Book-length treatments are now beginning to appear, including the edited collection by Finkenstadt, Held and Isham (2007), several chapters of Gelfand *et al.* (2009) and, most recently, Cressie and Wikle (2011).

Another important development, throughout the statistics discipline, has been the rise in popularity of R as a vehicle for the dissemination of new statistical methods through open-source software. Useful packages for the analysis of spatial point process data include `spatial`, `spatstat`, `MarkedPointProcess`, `splancs` and `spatialkernel`. All of these, and more, can be downloaded from the R project web-page, `www.r-project.org`. I predict with some confidence that the above list will be out-of-date by the time this appears in print.

Public-domain data-sets used in the book, and any errors of which I am aware, can be found on the book's web-page:

`http://wwww.lancs.ac.uk/staff/diggle/pointpatternbook`

My thanks are again due to many colleagues, in many places and over some forty years, who have provided me with such stimulating working environments, spanning the UK, Sweden, Australia and the USA. I was fortunate to begin my career under the wise guidance of the late Prof Robin Plackett at the University of Newcastle upon Tyne. Periods spent at the Royal College of Forestry Stockholm, CSIRO Australia and, most recently, the University of Liverpool, have taught me the inestimable value of working closely with subject-matter scientists. Visits to the Department of Biostatistics at Johns Hopkins University, Baltimore, stimulated an enduring interest in medical and public health applications. At Lancaster University, I have been privileged to work with a succession of talented young research students and staff, amongst whom special mention goes to Barry Rowlingson for his patient, if doomed, efforts over 25 years to teach me to compute efficiently.

Finally, my collaborators on the many jointly authored publications listed amongst the references should share the credit for whatever value the book may have, whereas responsibility for defects remains mine alone.

Peter J. Diggle, Lancaster

1

Introduction

CONTENTS

1.1 Spatial point patterns

Data in the form of a set of points, irregularly distributed within a region of space, arise in many different contexts; examples include locations of trees in a forest, of nests in a breeding colony of birds, or of nuclei in a microscopic section of tissue. We call any such data-set a *spatial point pattern* and refer to the locations as *events*, to distinguish these from arbitrary points of the region in question.

Figures 1.1 and 1.2 show two spatial point patterns in a square region. The first, due to Numata (1961), shows 65 Japanese black pine saplings in a square of side 5.7 metres whilst the second, extracted by Ripley (1977) from Strauss (1975), shows 62 redwood seedlings in a square of side 23 metres, approximately. The two patterns appear strikingly different. Figure 1.1 shows no obvious structure and might be regarded as a "completely random" pattern, in a sense that we shall define formally in due course. In Figure 1.2, on the other hand, the strong clustering of seedlings requires some biological explanation which, in this instance, is readily available. The seedlings cluster around redwood stumps, which are known to be present in the study region but whose locations have not been recorded. It is important to recognise that patterns like Figure 1.2 can arise either through some form of clustering mechanism or through environmental variation leading to local patches with relatively high concentrations of events. Here, as elsewhere, failure to record relevant biological information limits the conclusions which can be drawn from a statistical

1

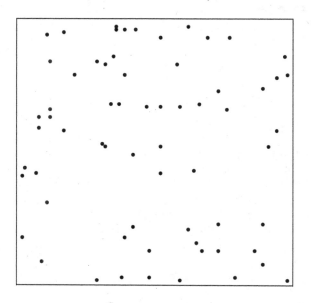

FIGURE 1.1
Locations of 65 Japanese black pine saplings in a square of side-length 5.7 metres (Numata, 1961).

analysis. We therefore describe patterns like Figure 1.2 as "aggregated" to avoid the mechanistic connotations of the perhaps more obvious term "clustered."

Figure 1.3 shows a further qualitatively different type of pattern, formed in this case by the centres of 42 biological cells (Crick and Lawrence, 1975; Ripley, 1977). The cell centres are distributed more or less regularly over the unit square, improbably so unless there is some associated regulating mechanism operating to encourage an even spatial distribution of cell centres. A possible explanation is that the cell centres are merely convenient reference points for cells whose physical size is non-negligible relative to the scale of observation. Quite generally, and again without wishing to imply any specific causal mechanism, we refer to such patterns as "regular."

The nature of the pattern generated by a biological process can be affected by the physical scale on which the process is observed. At a sufficiently large scale most natural environments exhibit heterogeneity, which will tend to produce aggregated patterns. At a smaller scale, environmental variation will be

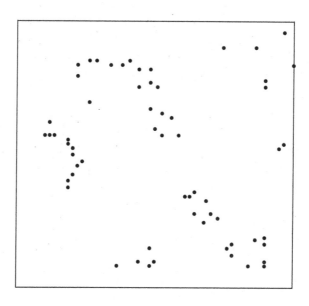

FIGURE 1.2

Locations of 62 redwood seedlings in a square of side-length 23 metres
(Strauss, 1975; Ripley, 1977).

less pronounced and the major determinant of pattern may then be the nature
of the interactions amongst the events themselves. For example, vegetative
propagation of individual shoots will tend to produce small-scale aggregation
whereas competition for space will encourage regularity. Our classification of
patterns as regular, random or aggregated is therefore an over-simplification,
but a useful one at an early stage of analysis. At a later stage, this simplistic
approach can be abandoned in favour of a more detailed, and essentially mul-
tidimensional, description of pattern that can be obtained either by the use of
a variety of functional summary statistics or by formulating an explicit model
of the underlying process. The approach taken in this book will be to develop
methods for the analysis of spatial patterns based on *stochastic models*, which
assume that the events are generated by some underlying random mechanism.

Our fourth example, shown in Figure 1.4, introduces the idea of a *multivariate* point pattern. In this example, the points represent cells of two different
types (hence, *bivariate*) in the retina of a rabbit. The data consist of the locations of 294 displaced amacrine cells, amongst which 152 are of a type that

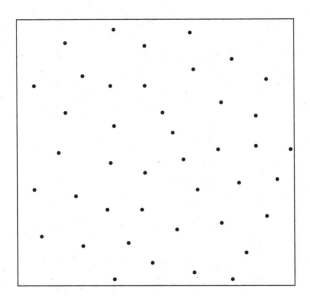

FIGURE 1.3
Locations of 42 cell centres in a unit square (Ripley, 1977).

transmits information to the brain when a light goes *on*, whilst the remaining 142 transmit information when a light goes *off*. The relationship between the two component patterns can help to explain the developmental processes that operate within the immature retina. We shall re-examine the data from this point of view in Section 4.7.

Our fifth example is of a *spatio-temporal point pattern*, in which the data provide both the location and the time of occurrence of events of scientific interest within a specified spatial region and time-interval. Figure 1.5 shows the residential locations and dates of 100 consecutive cases of non-specific gastrointestinal symptoms, as reported between 1 and 8 January 2001 to NHS Direct, a 24-hour phone-based triage service operated by the UK National Health Service, by residents in the county of Hampshire. The cases naturally cluster in areas of relatively high population density, but there is at least a hint that cases close in time (circles of the same radius of nearly so) are also closer spatially than might be expected by chance. If true, this would suggest that multiple cases may be the result of infections from a common source.

Spatio-temporal patterns are better examined dynamically than statically.

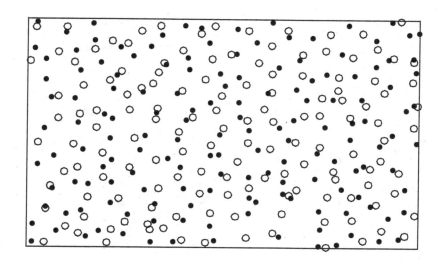

FIGURE 1.4
Locations of 294 displaced amacrine cells in the retina of a rabbit. Solid and open circles respectively identify *on* and *off* cells.

The data shown in Figure 1.5 are a sub-set of a much larger data-set reported in Diggle *et al.* (2003); an animation of the complete data-set by Barry Rowlingson can be viewed from the book's web-site.

We shall assume throughout this book that the spatial region of interest is essentially planar, although most of the ideas extend, at least in principle, to other dimensions. Even in one dimension, the distinction between temporal and spatial point patterns is important. In the case of series of events irregularly distributed in time, for example division times in a cell proliferation process, stochastic models and their associated statistical methods reflect the essentially unidirectional quality of the time dimension, whereas in the corresponding spatial case, for example nesting sites along the bank of a canal, no such directionality exists. Cox and Lewis (1966) give an excellent introduction to the analysis of temporal point patterns, whilst Daley and Vere-Jones (2002, 2005) discuss the underlying point process theory in depth.

All of our examples involve applications in the life sciences, although similar problems arise in many other disciplines. For examples in archaeology, astronomy and geography see, respectively, Hodder and Orton (1976), Peebles (1974) and Cliff and Ord (1981). To some extent, the methods that we

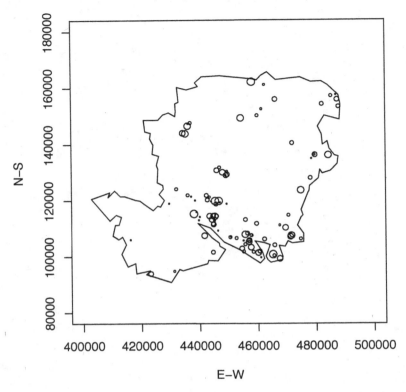

FIGURE 1.5
Locations of cases of non-specific gastrointestinal symptoms reported to NHS
Direct Hampshire, UK, between 1 January and 8 January 2001. The radius of
each plotted circle codifies the reporting date (smallest for 1 January, largest
for 8 January).

describe remain useful (and have certainly been used) in these other areas of
application, but should not be adopted uncritically. In particular, our stochastic models will be motivated by simple considerations of possible underlying
biological mechanisms that may or may not be relevant in other disciplines.

1.2 Sampling

The selection of the study region, A say, merits some discussion. In some
applications, A is objectively determined by the problem in hand, and inferences are required in terms of a process defined on A itself. One example of

this would be a map of all nesting sites on an island. More commonly, A is selected from some much larger region. The selection of A may then be made according to a probability sampling scheme, or it may simply reflect the experimenter's view that A is in some sense representative of the larger region. In either case, but particularly the latter, inferences drawn from an analysis will carry much greater conviction if consistency over replicate data-sets can be demonstrated.

As an alternative to intensive mapping within a single region A, the experimenter may choose to record limited information from a large number of smaller regions, for example the number of events in each region. In this context, the regions are called *quadrats* and the data are referred to as *quadrat counts*. A quadrat was originally a square of side-length 1 metre, used by the Uppsala school of plant ecologists as the basic sampling unit for investigating plant communities in the field (Du Rietz, 1929).

Quadrat sampling remains a popular field technique in plant ecology and elsewhere, but in some contexts it is rather impractical. This led to the development, initially in the American forestry literature (Cottam and Curtis, 1949), of a number of *distance methods* for sampling spatial point patterns. In these, the basic sampling unit is a point, and information is recorded in the form of distances to neighbouring events, for example the distances to the first few nearest events.

We shall refer to quadrat count and distance methods as *sparse sampling* methods, to distinguish them from intensive mapping exercises. The appropriate techniques for analysing data obtained by sparse sampling and by intensive mapping are quite different. Also, analyses of sparsely sampled patterns typically have more limited objectives than do analyses of mapped patterns.

A particular form of quadrat sampling, intermediate between sparse sampling and intensive mapping, consists of partitioning the study region into disjoint sub-regions and recording the number of events in each sub-region. Data of this kind arise in two very different ways. The first, which has a long tradition as a method of field sampling in plant ecology, is when a rectangular study region is partitioned into a regular grid of square or rectangular quadrats, and a count is taken within each quadrat. The second, which typically arises in environmental epidemiology, is when health outcome data are routinely maintained as counts of the numbers of events in administratively defined sub-regions. In either case, the resulting data can be represented as a realisation of a high-dimensional multivariate random variable, $Y = (Y_1, ..., Y_n)$ say, where Y_i denotes the number of events in the ith sub-region. In this setting, a stochastic model for the underlying point process would induce a unique statistical distribution on Y, but in practice the form of this distribution tends to be intractable except in a few special cases. A pragmatic alternative is to formulate a model directly for the distribution of Y, without reference to any underlying point process. The usual method of construction is to specify the set of conditional distributions of each Y_i given all other Y_j. Models of this kind are called *Markov random fields*. Their construction must satisfy sometimes non-obvious constraints to ensure self-consistency. Besag (1974) is an early, and very influential, account; see also Rue and Held (2005) or the relevant chapters of Gelfand *et al.* (2009) for detailed accounts.

Replicated sampling of mapped patterns is surprisingly rare. Ecological investigations have certainly compared patterns in study regions deliberately selected to represent different environmental conditions (see, for example, Bagchi *et al.* 2011), but I am not aware of corresponding studies which have been designed with a view to establishing the consistency of patterns in ostensibly similar regions. Pseudo-replication can always be achieved by partitioning a single region into two or more sub-regions. Genuine replication is more common in fields such as neuroanatomy, where the natural spatial sampling unit is a single field of view under a microscope, and there is a well-established tradition of using hierarchical sampling designs of the form: multiple fields of view within tissue sections; multiple tissue sections within subjects; multiple subjects within experimental treatment groups. Studies of this kind lend themselves to design-based inference as an alternative to the more widely prevailing model-based inference for spatial point patterns. See, for example, Diggle, Lange and Benes (1991), Baddeley *et al.* (1993) or Eglen, Diggle and Troy (2005).

1.3   Edge-effects

Edge-effects arise in spatial point pattern analysis when, as is often the case in practice, the region A on which the pattern is observed is part of a larger region on which the underlying process operates. The essential difficulty is then that unobserved events outside A may interact with observed events within A but, precisely because the events in question are not observed, it is difficult to take proper account of this.

For some kinds of exploratory analysis, edge-effects can safely be ignored. We shall discuss when and why this is so at appropriate points in the text. More generally, we can distinguish between three broad approaches to handling edge-effects: the use of buffer zones; explicit adjustments to take account of unobserved events; and, when A is rectangular, wrapping A onto a torus by identifying opposite edges. We will illustrate each of these approaches by considering a statistic that arises in several contexts, namely the number of events that occur within a specified distance of an arbitrary event or location.

The buffer zone method consists of carrying out all aspects of the statistical analysis after conditioning on the locations of all events which fall within a buffer zone B consisting of all points less than a specified distance, d_0 say, from the edge of A. Let $C = A - B$ denote the remainder of A after subtracting the buffer zone. Then, it is clear that for any event or location $x \in C$, the *observed* number of events within a distance d of x must equal the actual number of events *in the underlying process* within distance d of x, provided $d \leq d_0$, whereas for $d > d_0$ the observed number may be less than the actual number, thereby biasing any estimates based on these observed numbers. The choice of d_0 in the buffer zone method is awkward, since too small a value leaves residual edge-effects, whereas too large a value effectively throws away data unnecessarily. However, the method can be applied in adaptive form, varying the value of d_0 according to the particular statistic being used.

The adjustment method operates by making an "on average" adjustment for the unobserved events outside A. Again using our simple example to illustrate, if we count the observed number, n say, of events within distance d of a location x, and $a(d)$ denotes the area of intersection between A and a disc of radius d centred on the location x, then an intuitively sensible estimate of the actual number of events within distance d of x is $n\pi d^2/a(d)$. The adjustment method is attractive because it makes full use of the observed data, especially when relatively large-scale effects are of interest. Note, however, that the adjustments typically lead to an increased sampling variance for the edge-adjusted estimator by comparison with its unadjusted counterpart. In essence, this is an example of the common trade-off in statistical estimation between bias and variance, as edge-corrections seek to eliminate bias at the expense of some increase in variance.

Toroidal wrapping of a rectangular A is not so much an edge-correction method as a trick to eliminate edge-effects in particular circumstances. Its most common use is as a convenient way of simulating realisations of various kinds of point process. For example, suppose that we wish to simulate a point process model for the cell centre data shown in Figure 1.3, the most obvious feature of which is that no two events can occur too close together. If we attempt to simulate a process of this kind directly on a unit square A, then points near the edge of A are favoured over points near the centre of A as potential locations, because of the absence of potentially inhibiting events outside A. By simulating the process on a torus and subsequently unwrapping to a unit square for presentation, we avoid this effect. Note that it will seldom make sense to wrap observed data onto a torus for analysis; for example, if we were to do this with the cell centre data, then we would observe some very small toroidal distances between pairs of cells, thus distorting the inhibitory nature of the underlying process.

1.4 Complete spatial randomness

The hypothesis of *complete spatial randomness* (henceforth CSR) for a spatial point pattern asserts that (i) the number of events in any planar region A with area $|A|$ follows a Poisson distribution with mean $\lambda|A|$; (ii) given n events x_i in a region A, the x_i are an independent random sample from the uniform dis-

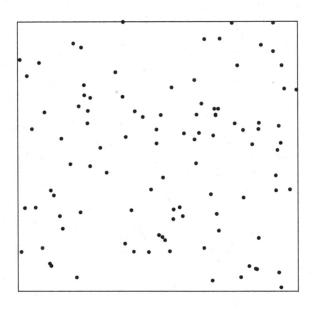

FIGURE 1.6
Realisation of CSR: 100 events in a unit square.

tribution on A. The self-consistency of (i) and (ii) is not immediately obvious, but will be established in Chapter 4. In (i), the constant λ is the *intensity*, or mean number of events per unit area. According to (i), CSR therefore implies that the intensity of events does not vary over the plane. According to (ii), CSR also implies that there are no interactions amongst the events. For example, the independence assumption in (ii) would be violated if the existence of an event at x either encouraged or inhibited the occurrence of other events in the neighbourhood of x. In developing tests of CSR for sparsely sampled patterns the starting point will be property (i), whilst for mapped patterns it is more usual to start with (ii), i.e. to analyse the pattern conditional on the observed number of events.

Intuitive ideas about what constitutes a "random pattern" can be misleading. Figure 1.6 shows a realization of 100 events independently and uniformly distributed on the unit square. Any visual impression of aggregation is illusory. Note also the superficial similarity to Figure 1.1.

Our interest in CSR is that it represents an idealized standard which, if strictly unattainable in practice, may nevertheless be tenable as a convenient first approximation. Most analyses begin with a test of CSR, and there are several good reasons for this. Firstly, a pattern for which CSR is not rejected scarcely merits any further formal statistical analysis. Secondly, tests are used as a means of exploring a set of data, rather than because rejection of CSR is of intrinsic interest. Greig-Smith, in the discussion of Bartlett (1971), emphasized that ecologists often know CSR to be untenable but nevertheless use tests of CSR as aids to the formulation of ecologically interesting hypotheses concerning pattern and its genesis. Thirdly, CSR acts as a dividing hypothesis to distinguish between patterns which are broadly classifiable as "regular" or "aggregated".

Another use of CSR is as a building block in the construction of more complex models. We shall return to this topic in Chapter 6.

1.5 Objectives of statistical analysis

In any particular application, the objectives of a statistical analysis should be determined by the scientist's objectives in collecting the data in question. We have already given reasons for beginning an analysis with a test of CSR. What to do next will vary according to context.

In sparse sampling exercises, a specific objective may be to estimate the intensity. For example, in forestry surveys an important quantity to be estimated is the "stocking density," or number of stems per hectare. The nature of the pattern might then be of interest only in so much as it affects the sampling distribution of the estimator.

With mapped data, the scientist will usually want a more detailed description of the observed pattern than can be provided by a test of CSR. One way to achieve this is by formulating a parametric stochastic model and fitting it to the data. If a model can be found that fits the data well, the estimated values of its parameters provide summary statistics that can be used to compare ostensibly similar data-sets. More ambitiously, a fitted model can provide an explanation of the underlying scientific processes. But this must involve an element of *non-statistical* inference: quite apart from the obvious fact that a model which fits the data is not necessarily correct in any absolute sense, we shall see in Chapter 4 that a simple stochastic model for a spatial point pattern may admit more than one scientific interpretation.

Of course, it is generally the case that modelling itself is only a means to a wider end. A well-formulated, and well-fitting, model provides a parsimonious description of a complex pattern, and one which will be especially useful if its parameters can be related to scientific hypotheses about the underlying phenomenon being studied.

Model-fitting is particularly difficult for very heterogeneous data-sets. In such cases, it may be unhelpful to force a parametric analysis based on untenable assumptions. For example, a generic problem in environmental epidemiology is to estimate the spatial variation in the risk of a particular disease, using data on the locations of individual cases in a geographical region. One approach to this problem might be to formulate an idealised model for the observed spatial pattern of cases under the assumption that risk is spatially constant, and to investigate deviations of the observed pattern from this model. An alternative, which would be more in line with classical epidemiological methods, would be to make a nonparametric comparison between the pattern of cases and a second pattern of *controls*, defined to be a random sample from the population at risk. We shall discuss this idea in detail in Chapter 9.

1.6   The Dirichlet tessellation

Given n distinct events x_i in a planar region A, we can assign to x_i a "territory" consisting of that part of A which is closer to x_i than to any other x_j. This construction, referred to either as the *Dirichlet tessellation* or *Voronoi tessellation* of the events in A, has been incorporated into stochastic models of natural phenomena such as inter-plant competition. In these models, plants whose territories abut are assumed to be in direct competition for available nutrient; see, for example, Cormack (1979, pp. 171–175). For large n, the tessellation is also the basis of a computationally efficient solution to a number of problems involving the calculation of distances between events.

Except possibly along the boundary of A, each territory or *cell* is a convex polygonal region. Events x_i and x_j whose cells share a common boundary segment are said to be *contiguous*. Typically, each cell vertex is common to three cells, and the lines joining the pairs of contiguous events define a triangulation of the x_i, called the *Delaunay triangulation*. Thus, cell boundaries can be obtained as the perpendicular bisectors of the edges of the triangulation, and cell vertices are the corresponding circumcentres. Figure 1.7 shows a simple example of both the tessellation and the triangulation associated with $n = 12$ events in a unit square. Rogers (1964) discusses the mathematical properties of the Dirichlet tessellation in a general p-dimensional setting.

The construction of the Dirichlet tessellation and the associated Delaunay triangulation rapidly becomes a non-trivial exercise as n increases. Green and Sibson (1978) give a remarkably efficient algorithm whose computational cost increases roughly as $n^{1.5}$.

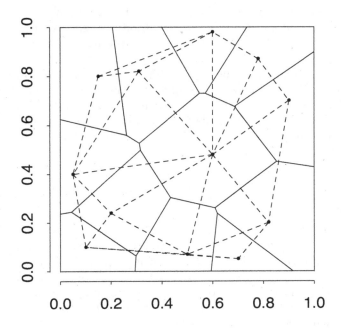

FIGURE 1.7
The Dirichlet tessellation (——) and Delaunay triangulation ($---$) associated with 12 points in a unit square.

1.7 Monte Carlo tests

Even simple stochastic models for spatial point patterns lead to intractable distribution theory, and in order to test models against data we shall make extensive use of Monte Carlo tests (Barnard, 1963).

Quite generally, let u_1 be the observed value of a statistic U and let $u_i : i = 2, ..., s$ be corresponding values generated by independent random sampling from the distribution of U under a simple hypothesis $\mathcal{H}$. Let $u_{(j)}$ denote the jth largest amongst $u_i : i = 1, ..., s$ Then, under $\mathcal{H}$,

$$P\{u_1 = u_{(j)}\} = s^{-1}, \ j = 1, ..., s,$$

and rejection of $\mathcal{H}$ on the basis that u_1 ranks kth largest or higher gives an exact, one-sided test of size k/s. This assumes that the values of the u_i are all different, so that the ranking of u_1 is unambiguous. If U is a discrete random variable, for example a count, tied values are possible and we then adopt the conservative rule of choosing the least extreme rank for u_1. The extension to two-sided tests is clear.

Hope (1968) gives a number of examples to show that the loss of power resulting from a Monte Carlo implementation is slight, so that s need not be very large. For a one-sided test at the conventional 5% level, $s = 100$ is adequate.

Power loss is related to Marriott's (1979) investigation of "blurred critical regions", which arise because a value of u_1 which would be declared significant in a classical test may not be declared significant in a Monte Carlo test, and *vice versa*. Let the (unknown) distribution function of U under $\mathcal{H}$ be $F(u)$. For a one-sided 5% test with $s = 20k$, the probability that we reject $\mathcal{H}$, given that $U = u_1$, is

$$p(u_1) = \sum_{r=0}^{k-1} \binom{s-1}{r} \{1 - F(u_1)\}^r \{F(u_1)\}^{s-1-r}. \qquad (1.1)$$

For a classical test, represented here by the limit $s \to \infty$, $p(u_1)$ is 1 or 0 accordingly as $F(u_1)$ is greater or less than 0.95. The effect of the "blurring" introduced by (1.1) is measured by Marriott's Table 1, here reproduced as Figure 1.8. Marriott concludes also that the extent of the blurring depends primarily on k, so that if $s = 100$ is judged to be acceptable for a test at the 5% level, then $s = 500$ should be used for a test at the 1% level, and *pro rata* for tests at smaller levels. These recommended values of s are smaller than would be considered adequate for the estimation of $P\{U > u_1 \mid \mathcal{H}\}$. This is essentially a consequence of the blurring effect. It is the *rank* of u_1, and not u_1 itself, which is the test statistic. A Monte Carlo test may fail to reject $\mathcal{H}$ when a classical test would have done so, representing a loss of power, but may also reject $\mathcal{H}$ when a classical test would have done so, representing a gain in power.

A technical point concerning the use of Monte Carlo tests is that in practice random sampling will be replaced by pseudo-random sampling. In the past, we have used the generator supplied in the NAG (1977) subroutine library or a Fortran implementation of the Wichmann and Hill (1982) generator, but we now routinely use the built-in R function `runif()` (Venables and Ripley, 1994). Other references for users who wish to understand the theory underlying pseudo-random number generators include Kennedy and Gentle (1980) or Ripley (1987).

A more subtle, but potentially more important, criticism is that Monte Carlo tests encourage "data-dredging", since the user can choose the statistic U to focus on any seemingly aberrant feature of their data. Whilst we admit that this is a danger, it should be obvious that "significant" results based on pathological test statistics are of no practical value.

An inherent weakness of the Monte Carlo approach is its restriction to simple hypotheses $\mathcal{H}$. Composite hypotheses can be tested if pseudo-random sampling is made conditional on the observed values of sufficient statistics for any unknown parameters, but this is seldom feasible. Note that a goodness-of-fit test that ignores the effects of estimating parameters will tend to be

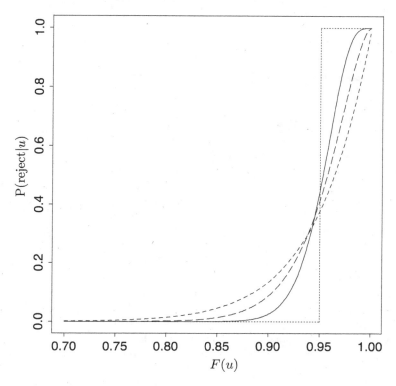

FIGURE 1.8
Blurred critical regions for one-sided, 5% Monte Carlo tests with $s = 20$, 40, 100 and $s \to \infty$ (adapted from Marriott, 1979).

conservative. This particular difficulty does not arise with tests of CSR for mapped data, because the observed number of events n is sufficient for the intensity λ, and conditional on n CSR is a simple hypothesis. But it does affect the assessment of goodness-of-fit for more general stochastic models. An approximate remedy, which we discuss further in Chapter 6, is to measure goodness-of-fit by a statistic that is not directly related to the procedure used to estimate the parameters of the model.

The principal advantage to be set against the above is that the investigator need not be constrained by known distribution theory, but rather can and should use informative statistics of their own choosing.

When asymptotic distribution theory is available, Monte Carlo testing provides an exact alternative for small samples and a useful check on the applicability of the asymptotic theory. If the results of classical and Monte Carlo tests are in substantial agreement, little or nothing has been lost; if not, the explanation is usually that the classical test uses inappropriate distributional assumptions.

1.8   Software

Spatial point pattern analysis is computationally intensive, not least because of the heavy reliance on Monte Carlo methods of inference. As noted in the Preface, R has become the computing environment of choice for many statisticians. The splancs package (Rowlingson and Diggle, 1993) gives a wide range of functions for statistical analysis of spatial point patterns. The Spatstat library, written by Adrian Baddeley and Rolf Turner, also implements a wide range of methods, with a stronger emphasis than splancs on parametric modelling. Many of the analyses reported in this book were implemented using a combination of splancs, Spatstat and Voronoi (a package for computation of the Dirichlet tessellation, written by Rolf Turner), together with some additional functions written by the author.

More sophisticated displays than those shown in this book, for example colour-coded overlays of point pattern maps and contour maps, can most easily be produced using a Geographical Information System (GIS). A wide variety of commercial and open-source GIS packages are now available. Also, a number of R packages have been written to provide GIS-like functionality within the R environment. For a detailed description of spatial data-handling in R, see for example Bivand, Pebesma and Gomez-Rubio (2008).

2

Preliminary testing

CONTENTS

2.1 Tests of complete spatial randomness

Although complete spatial randomness is of limited scientific interest in itself, there are several good reasons why we might begin an analysis with a test of CSR: rejection of CSR is a minimal prerequisite to any serious attempt to model an observed pattern; tests are used to explore a set of data and to assist in the formulation of plausible alternatives to CSR; CSR operates as a dividing hypothesis between regular and aggregated patterns.

In view of the above, the present discussion emphasizes two aspects: the value of graphical methods, which will almost always be informative and will sometimes make formal testing unnecessary; and informal combination of several complementary tests, to indicate the nature of any departure from CSR. With regard to the second of these, if a single assessment of significance is required the following result is useful. Suppose that the attained significance levels of k, not necessarily independent tests of CSR, are $p_j : j = 1, ..., k$ and let p_{min} be the smallest such p_j, corresponding to the most significant departure from CSR. Then, under CSR,

$$p \leq P\{p_{min} \leq p\} \leq kp. \tag{2.1}$$

For k independent tests, the exact result is

$$P\{p_{min} \leq p\} = 1 - (1 - p)^k.$$

Cox (1977) points out that using multiple tests as part of a diagnostic procedure makes practical sense only if the various tests examine different aspects of pattern, so that a significant result for one test does not prevent a sensible interpretation of the others.

We acknowledge that testing complete spatial randomness is a very unambitious agenda in itself, and should be seen as no more than a natural starting point. From a pedagogical point of view, it provides a historical perspective on the early development of the subject, and an opportunity to illustrate a number of general issues in the simplest possible setting. These include the role of Monte Carlo methods, the need to assess the relative merits of intuitively sensible but *ad hoc* methods and, perhaps most importantly, the need to take account of the inherent dependence amongst multiple measurements derived from a single point pattern. In the remainder of the chapter we will therefore describe a number of different tests that have been proposed, and assess their strengths and weaknesses. As illustrative examples, we shall use repeatedly the three data-sets shown in Figures 1.1 to 1.3, each of which has a straightforward interpretation. More ambitious analyses of these and other data-sets will appear in later chapters, when we discuss the formulation and fitting of stochastic models other than CSR.

2.2   Inter-event distances

One possible summary description of a pattern of n events in a region A is the empirical distribution of the $\frac{1}{2}n(n-1)$ inter-event distances, t_{ij} say. The corresponding theoretical distribution of the distance T between two events independently and uniformly distributed in A depends on the size and shape of A, but is expressible in closed form for the most common cases of square or circular A (Bartlett, 1964).

For A a square of unit side, the distribution function of T is

$$H(t) = \begin{cases} \pi t^2 - 8t^3/3 + t^4/2 & : \ 0 \le t \le 1 \\ 1/3 - 2t^2 - t^4/2 + 4(t^2 - 1)^{\frac{1}{2}}(2t^2 + 1)/3 & \\ \quad + 2t^2 \sin^{-1}(2t^{-2} - 1) & : \ 1 < t \le \sqrt{2} \end{cases} \tag{2.2}$$

whilst for a circle of unit radius the corresponding expression is

$$H(t) = 1 + \pi^{-1}\{2(t^2 - 1)\cos^{-1}(t/2) - t(1 + t^2/2)\sqrt{(1 - t^2/4)}\} \tag{2.3}$$

for all $0 \le t \le 2$.

We now develop a test of CSR based specifically on inter-event distances; the general approach is applicable to other summary descriptions and will reappear in later sections.

Assume that for the particular region A in question, $H(t)$ is known. Calculate the empirical distribution function (henceforth abbreviated to EDF) of inter-event distances. This function, $\hat{H}_1(t)$ say, represents the observed proportion of inter-event distances t_{ij} which are at most t; thus,

$$\hat{H}_1(t) = \frac{1}{2}n(n - 1)\}^{-1}\#(t_{ij} \le t),$$

where $\#$ means "the number of." Now prepare a plot of $\hat{H}_1(t)$ as ordinate against $H(t)$ as abscissa. If the data are compatible with CSR, the plot should be roughly linear. To assess the significance or otherwise of departures from linearity, the conventional approach would be to find the sampling distribution of $\hat{H}_1(t)$ under CSR. But this is complicated by the dependence between inter-event distances with a common end-point, and we therefore proceed as follows. Calculate EDF's $\hat{H}_i(t), i = 2, 3, ..., s$, from each of $s - 1$ independent simulations of n events independently and uniformly distributed on A, and define *upper* and *lower simulation envelopes*,

$$U(t) = \max\{\hat{H}_i(t)\}; \quad L(t) = \min\{\hat{H}_i(t)\}, \tag{2.4}$$

where in each case, i runs from 2 to s. These simulation envelopes can also be plotted against $H(t)$, and have the property that under CSR, and for each t,

$$P\{\hat{H}_1(t) > U(t)\} = P\{\hat{H}_1(t) < L(t)\} = s^{-1}. \tag{2.5}$$

Simulation envelopes are intended to help in the interpretation of the plot of $\hat{H}_1(t)$ against $H(t)$, and we shall shortly give examples of their use. Two of the many possible approaches to the construction of an exact Monte Carlo test of CSR are as follows.

(i) Choose t_0 and define $u_i = \hat{H}_i(t_0)$. As described in Section 1.7, the rank of u_1, amongst the u_i provides the basis of a test because under CSR all rankings of u_1 are equi-probable.

(ii)   Define u_i to be a measure of the discrepancy between $\hat{H}_i(t)$ and $H(t)$ over the whole range of t, for example

$$u_i = \int \{\hat{H}_i(t) - H(t)\}^2 dt \qquad (2.6)$$

and again proceed to a test based on the rank of u_1.

The first approach makes sense only if t_0 can be chosen in a way that is natural to the problem in hand. The second has the advantage of objectivity but we shall see that in the particular context of inter-event distances it often gives a very weak test. In any event, no single test statistic should be allowed to over-ride a critical inspection of the EDF plot.

If the region A is one for which the theoretical distribution function $H(t)$ is unknown, a test can still be carried out if, in (2.6), $H(t)$ is replaced by

$$\bar{H}_i(t) = (s-1)^{-1} \sum_{j \neq i} \hat{H}_j(t).$$

The u_i are no longer independent under CSR, but they are exchangeable and the required property that under CSR all rankings of u_1 are equi-probable therefore still holds. Similarly, the graphical procedure then consists of plotting $\hat{H}_1(t), U(t)$ and $L(t)$ against $\bar{H}_1(t)$. Note that because $\bar{H}_1(t)$ involves only the simulations of CSR and not the original data, it provides an unbiased estimate of $H(t)$ under the null hypothesis.

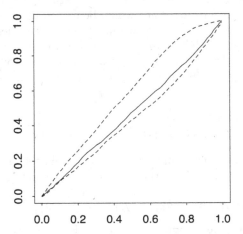

FIGURE 2.1
EDF plot of inter-event distances for Japanese black pine saplings. —— : data; − − − : upper and lower envelopes from 99 simulations of CSR.

2.2.1 Analysis of Japanese black pine saplings

Figure 2.1 shows the plot of $\hat{H}_1(t)$, $U(t)$ and $L(t)$ against $H(t)$ for Numata's data previously given as Figure 1.1. Note that $\hat{H}_1(t)$ lies close to $H(t)$ and between $U(t)$ and $L(t)$ throughout its range, which suggests acceptance of CSR. A formal test based on the integrated squared difference (2.6) and 99 simulations ($s = 100$) leads to an attained significance level of 0.37, and we conclude that these data are compatible with a completely random spatial distribution of saplings over the study region. The same conclusion was reached by Bartlett (1964) and by Besag and Diggle (1977), who based their test on Pearson's X^2 goodness-of-fit statistic applied to a histogram of inter-event distances.

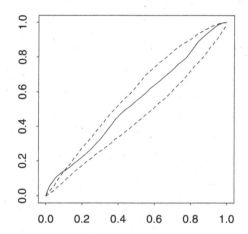

FIGURE 2.2
EDF plot of inter-event distances for redwood seedlings. ——— : data; – – – : upper and lower envelopes from 99 simulations of CSR.

2.2.2 Analysis of redwood seedlings

For the redwood data of Figure 1.2, a test based on (2.6) again suggests acceptance of CSR with an attained significance level of 0.22, but a detailed inspection of the EDF plot in Figure 2.2 leads to a different conclusion. We see that $\hat{H}_1(t)$ is greater than $H(t)$ throughout its range and in particular is greater than $U(t)$ for both very small and very large values of $H(t)$. The excess of small inter-event distances is compatible with an underlying clustering mechanism for which, as we have seen, there is a ready biological explanation. Further reinforcement of this conclusion, if any were needed, lies in Strauss's remark that a distance of 6 feet (approximately 2 metres) on the ground, corresponding to $t \approx 0.08$, "was thought to be very roughly the range at which a pair of seedlings could interact." This suggests that $\hat{H}_1(0.08)$, the observed proportion of inter-event distances less than or equal to 6 feet, is a reasonable test statistic. Since $\hat{H}_1(0.08) > U(0.08)$, it follows that CSR is rejected at a (one-sided) attained significance level of 0.01. Strictly, of course, this conclusion would only be valid if we had chosen our test statistic to be $\hat{H}_1(0.08)$ before inspecting Figure 2.2. Perhaps a more useful message from this example is that it reinforces the value of looking at the EDF plot in conjunction with its simulation envelopes, rather than relying on the result of a formal test of significance.

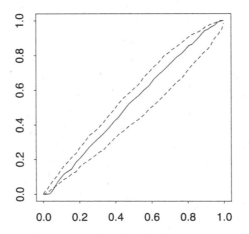

FIGURE 2.3
EDF plot of inter-event distances for biological cells. ——— : data; − − − : upper and lower envelopes from 99 simulations of CSR.

2.2.3   Analysis of biological cells

For Ripley's cell data previously given as Figure 1.3, a test based on (2.6) again suggests acceptance of CSR, this time with an attained significance level of 0.23, but inspection of the EDF plot in Figure 2.3 again suggests otherwise. The most striking feature of this plot is the complete absence of small inter-event distances, so that $\hat{H}_1(t) = 0$ for small t. This provides an explanation for the regular appearance of the observed pattern. Also, at large values of $H(t)$ we see that $\hat{H}_1(t)$ lies close to, albeit below, $U(t)$. This is unusual for a regular pattern, and encourages us to re-examine Figure 1.3. With the benefit of hindsight, we see a surprising lack of events in the corners of the unit square. This suggests that there may have been some difficulty in determining the boundary of the study region, which in turn would have led to empty spaces near the corners of the square region and a spurious deficit of large inter-event distances.

2.2.4   Small distances

For the redwood seedlings and the biological cells, the evidence against CSR derives from the excess or deficiency, respectively, of *small* inter-event distances. The main body of the distribution of inter-event distance is relatively insensitive to changes in pattern.

It follows that at larger values of t, departures from the null form of $H(t)$ are usually swamped by sampling fluctuations in $\hat{H}_1(t)$, unless n is very large. Thus, while the EDF plot is informative, the test based on (2.6) is not recommended.

An extreme case of concentrating on small inter-event distances would be to use as test statistic the minimum inter-event distance. This is theoretically attractive if regular alternatives to CSR are suspected, because for particular kinds of regular alternative to CSR it can be derived as a likelihood ratio statistic. Furthermore, an approximate test can be implemented without simulations. Silverman and Brown (1978) express the asymptotic null distribution of T_k, the kth smallest inter-event distance, as

$$n(n-1)\pi|A|^{-1}T_k^2 \sim \chi_{2k}^2. \tag{2.7}$$

Ripley and Silverman (1978) suggest that the chi-squared approximation is adequate for $k \leq 9$ when $n \geq 30$.

For the biological cells, the observed value of T_1, the minimum inter-event distance, is 0.086. With $n = 42$ and $|A| = 1$, (2.7) gives $P\{T_1 \geq 0.086\} < 0.001$ and CSR is conclusively rejected. The same test accepts CSR for both the Japanese black pines and the redwoods.

A disadvantage of this test for large data-sets is its sensitivity to recording inaccuracies. For example, suppose that n events in the unit square are mapped to an accuracy of two decimal places in each coordinate direction; this corresponds approximately to the accuracy achieved in Figures 1.1 to 1.3. Then, the observed value of T_1 must be either zero, and significantly small according to (2.7), or at least 0.01. From (2.7), we can deduce that the upper critical value of T_1 for a one-sided, 5%, test of CSR is approximately $1.38\{n(n-1)\}^{\frac{1}{2}}$, and this is *less* than 0.01 if $n \geq 139$, hence any non-zero value of T_1 would be declared significantly large. A test based on T_k for some value of $k > 1$ is more robust in this respect, but the choice of k then becomes rather arbitrary.

2.3   Nearest neighbour distances

For n events in a region A, let y_i denote the distance from the ith event to the nearest other event in A. The y_i are called *nearest neighbour distances*. Typically, the n nearest neighbour distances for a pattern of n events include duplicate measurements between reciprocal nearest neighbour pairs. We can calculate the EDF, $\hat{G}_1(y)$ say, of the nearest neighbour distances by analogy with the calculation used in Section 2.2 to obtain $\hat{H}_1(t)$. Thus,

$$\hat{G}_1(y) = n^{-1}\#(y_i \leq y).$$

In many practical situations, interactions between events exist, if at all, only at a small physical scale. For example, trees would be expected to compete for sunlight or nutrient within an area roughly confined to their crowns or root systems, respectively. In such cases, nearest neighbour distances provide an objective means of concentrating on "small" inter-event distances when a precise threshold distance cannot be specified in advance.

The theoretical distribution of the nearest neighbour distance Y under CSR depends on n and on A, and is not expressible in closed form because of complicated edge effects. An approximation that ignores these edge effects is obtained by noting that if $|A|$ denotes the area of A, then $\pi y^2 |A|^{-1}$ is the probability under CSR that an arbitrary event is within distance y of a specified event. Since the events are located independently, the approximate distribution function of Y is

$$G(y) = 1 - (1 - \pi y^2 |A|^{-1})^{n-1}.$$

A further approximation for large n, writing $\lambda = n|A|^{-1}$, is

$$G(y) = 1 - \exp(-\lambda \pi y^2) : y \geq 0. \tag{2.8}$$

This result is well known. In Chapter 3 we shall reach it by a different route, as a property of the homogeneous planar Poisson process.

The EDF $\hat{G}_1(y)$ can be compared with upper and lower simulation envelopes from simulated EDFs $\hat{G}_i(y) : i = 2, ..., s$ exactly as in Section 2.2. The approximate result (2.8) can be used to suggest a suitable range of tabulation but, because it is approximate, it is generally preferable to use the sample mean $\bar{G}_1(y)$ of simulated EDFs for linearisation of the EDF plot. Possible test statistics for a Monte Carlo test include the sample mean of the n nearest neighbour distances, or

$$u_i = \int \{\hat{G}_i(y) - \bar{G}_i(y)\}^2 dy, \tag{2.9}$$

where

$$\bar{G}_i(y) = (s-1)^{-1} \sum_{j \neq i} \hat{G}_j(y)$$

is defined by analogy with $\bar{H}_i(t)$ in Section 2.2. A test based on the sample mean, $\bar{y}$, was proposed by Clark and Evans (1954), but without proper allowance for the dependencies amongst the nearest neighbour distances. One possible advantage of a test based on $\bar{y}$ is that, as with the test based on the minimum inter-event distance, simulation is unnecessary. Donnelly (1978) has shown that, to a good approximation, the distribution of $\bar{y}$ under CSR is Normal, with mean and variance

$$E[\bar{y}] = 0.5(n^{-1}|A|)^{1/2} + (0.051 + 0.042n^{-1/2})n^{-1}P \qquad (2.10)$$

and

$$\text{Var}(\bar{y}) = 0.070n^{-2}|A| + 0.037(n^{-5}|A|)^{1/2}P \qquad (2.11)$$

where P denotes the perimeter length of A. Significantly small or large values of $\bar{y}$ indicate aggregation or regularity, respectively. A minor qualification is that these approximations break down for very convoluted regions A, in which case a Monte Carlo implementation is again necessary. Also, we maintain that informal inspection of the EDF plot is at least as important as formal testing.

The obvious method of computing nearest neighbour distances involves a crude search through all the inter-event distances. For sufficiently large n a more efficient method is to construct the Dirichlet tessellation of the n events, and then to search for nearest neighbour distances within the tessellation. This exploits the fact that, however large n is, each event is contiguous to, on average, six other events, one of which must be its nearest neighbour. As a result, only a small fraction of the inter-event distances need to be calculated. Peter Green (personal communication) has shown that the crude search is more efficient than the tessellation method for n less than about 500, but thereafter becomes progressively less efficient with increasing n.

2.3.1 Analysis of Japanese black pine saplings

Figure 2.4 shows the EDF plot of nearest neighbour distances for the Japanese black pine saplings, together with the upper and lower envelopes from 99 simulations of CSR. The plot suggests acceptance of CSR, as does a Monte Carlo test based on (2.9) with an attained significance level of 0.52. In addition the observed value of $\bar{y}$ is 0.0660, whilst (2.10) and (2.11) give $E[\bar{y}] = 0.0655$ and $\text{Var}(\bar{y}) = 0.000021$, and hence a standard Normal deviate of -0.11, again suggesting acceptance of CSR.

Incidentally, and in contrast to Figures 2.1, 2.2 and 2.3, the linear interpolation between values of the EDF calculated at intervals of 0.01 shows up clearly in Figure 2.4. However, the limited resolution of the data does not justify either a much finer tabulation or a more subtle interpolation rule.

2.3.2 Analysis of redwood seedlings

For the redwood seedlings, a Monte Carlo test based on (2.9) leads to emphatic rejection of CSR, with u_1 comfortably larger than all 99 simulated u_j.

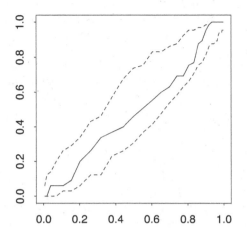

FIGURE 2.4
EDF plot of nearest neighbour distances for Japanese black pine saplings.
——— : data; − − − : upper and lower envelopes from 99 simulations of CSR.

The EDF plot, Figure 2.5, clearly shows the excess of small nearest neighbour distances which is a characteristic feature of aggregated patterns. The observed value of $\bar{y}$ is 0.0385. This corresponds to a standard Normal deviate of -5.96 and again provides strong evidence for rejection of CSR in favour of an aggregated alternative.

2.3.3   Analysis of biological cells

For the biological cells, (2.9) again gives a value of u_1 which is comfortably larger than all 99 simulated values u_j, whilst the EDF plot, Figure 2.6, now shows the deficiency of small nearest neighbour distances which is typical of regular patterns. For the Clark-Evans test, the observed value of $\bar{y}$ is 0.1283, the corresponding standard Normal deviate 6.30, and the conclusion emphatic rejection of CSR in favour of a regular alternative.

2.4   Point to nearest event distances

A related type of analysis uses distances x_i from each of m sample points in A to the nearest of the n events. The EDF $\hat{F}(x) = m^{-1}\#(x_i \leq x)$ measures the empty spaces in A, in the sense that $1 - \hat{F}(x)$ is an estimate of the area $|B(x)|$ of the region $B(x)$ consisting of all points in A a distance at least x

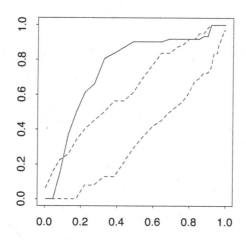

FIGURE 2.5

EDF plot of nearest neighbour distances for redwood seedlings. ——— : data;
− − − : upper and lower envelopes from 99 simulations of CSR.

from every one of the n events in A. The argument leading to (2.8) can be
repeated to show that, under CSR,

$$F(x) = 1 - \exp(-\pi\lambda x^2) : z \geq 0 \tag{2.12}$$

approximately, where $\lambda = n|A|^{-1}$.

Lotwick (1981) describes an algorithm, based on the Green-Sibson Dirich-
let tessellation algorithm, for computing $|B(x)|$ exactly when A is a rectangle.
In practice, using m points in a regular $k \times k$ grid gives an adequate approximation if k is reasonably large. A sensible choice for k depends to some extent
on the precise configuration of the n events in A and on the subsequent use
to which the estimator will be put. Diggle and Matérn (1981) recommend
$k \approx \sqrt{n}$ for estimating an unknown $F(x)$ from simulated realisations of a
point process, in which context we have the freedom to choose both the number of sample points per realisation and the number of realisations. Figure
2.7 shows, for the biological cell data, the degree of approximation introduced
by using $k = 7 \approx \sqrt{42}$ or $k = 14$. In a modern computing context, concern
about the computational effort of calculating $\hat{F}(x)$ is unnecessary, and there
is certainly no good statistical reason to limit the choice of k. However, it
is worth remembering that whilst large k will produce a very smooth curve
$\hat{F}(x)$, its statistical precision is still limited by n, the number of events.

By analogy with the procedure adopted for nearest neighbour distances, a
Monte Carlo test of CSR can be based on the statistic

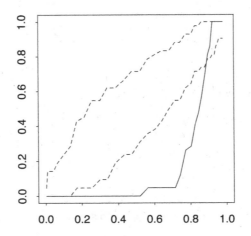

FIGURE 2.6
EDF plot of nearest neighbour distances for biological cells. ——— : data; — — —
: upper and lower envelopes from 99 simulations of CSR.

$$u_i = \int \{\hat{F}_i(x) - \bar{F}_i(x)\}^2 dx. \tag{2.13}$$

2.4.1 Analysis of Japanese black pine seedlings

Figure 2.8 shows the EDF plot for a point to nearest event analysis of Numata's data, using $k = 16$. We see that $\hat{F}_1(x)$ lies between the simulation envelopes and close to $\bar{F}_1(x)$ throughout its range. As in our previous analyses of these data, CSR is accepted.

2.4.2 Analysis of redwood seedlings

Figure 2.9 shows the corresponding EDF plot for the redwood data, again using $k = 16$. Now, $\hat{F}_1(x)$ lies below the lower simulation envelope for most of its range, and (2.13) leads to rejection of CSR with u_1 larger than all 99 simulated u_j. Note that $\hat{F}_1(x)$ drifts below the lower simulation envelope. This is typical of an aggregated pattern, and contrasts with the behaviour of $\hat{G}_1(y)$ for these data shown in Figure 2.5.

2.4.3 Analysis of biological cells

Figure 2.10 shows the comparable analysis of the biological cells, using $k = 14$. A test based on (2.13) again leads to rejection of CSR with an attained signif-

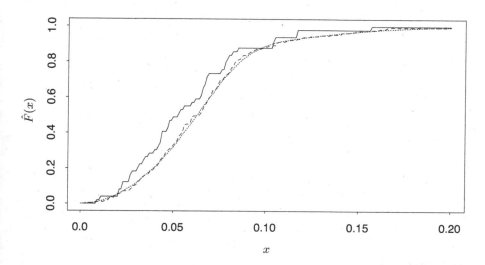

FIGURE 2.7
Calculation of $\hat{F}(x)$ for the biological cells data, using a $k \times k$ grid of sample points for different values of k. —— : $k = 7$; $- - -$: $k = 14$; .... : $k = 96$.

icance level of 0.02. The position of $\hat{F}_1(x)$ near or above the upper simulation envelope typifies a regular pattern and again contrasts with the behaviour of $\hat{G}_1(y)$ for these data in Figure 2.6.

2.5 Quadrat counts

An alternative to a distance-based approach is to partition A into m sub-regions, or quadrats, of equal area and to use the counts of numbers of events in the m quadrats to test CSR. The choice of sub-regions for this exercise is somewhat arbitrary. For ease of presentation, and because it represents common practice, we shall assume that A is the unit square and is partitioned into a regular $k \times k$ grid of square sub-regions, so that $m = k^2$. Let $n_i : i = 1, ..., m$ be the quadrat counts which result from this partitioning of A and write $\bar{n} = n/m$ for the sample mean of the n_i. An obvious statistic to test for departures from the uniform distribution on A implied by CSR is Pearson's criterion,

$$X^2 = \sum_{i=1}^{m} (n_i - \bar{n})^2 / \bar{n}, \qquad (2.14)$$

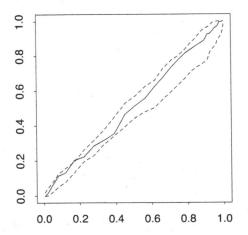

FIGURE 2.8
EDF plot of point to nearest event distances for Japanese black pine saplings.
——— : data; − − − : upper and lower envelopes from 99 simulations of CSR.

whose null distribution is χ^2_{m-1}, to a good approximation provided that $\bar{n}$ is not too small.

Note that X^2 is just $m-1$ times the sample variance-to-mean ratio of the observed quadrat counts, which Fisher *et al.* (1922) introduced in order to test the hypothesis that the counts follow a Poisson distribution. The relationship between a uniform distribution of events and a Poisson distribution of quadrat counts is not entirely transparent, but is implicit in our definition of CSR and will be discussed further in Chapter 4. Note also that in the present context the null hypothesis may fail either because of a non-uniform distribution of events in A or because of dependencies amongst the events. In particular, significantly large or small values of X^2 are both of interest, and respectively indicate a tendency towards an aggregated or a regular spatial distribution of events in A.

2.5.1   Analysis of Japanese black pine seedlings

For the 65 Japanese black pine saplings, the conservative rule that expected frequencies should be at least five suggests using a 3×3 grid. This gives an array of counts

$$
\begin{array}{ccc}
6 & 15 & 7 \\
10 & 4 & 3 \\
4 & 8 & 8
\end{array}
$$

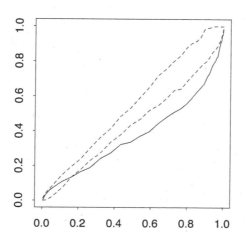

FIGURE 2.9
EDF plot of point to nearest event distances for redwood seedlings. ———— :
data; – – – : upper and lower envelopes from 99 simulations of CSR.

for which $X^2 = 15.17$, corresponding to a one-sided attained significance level
of $p = P(\chi_8^2 > 15.17) \approx 0.06$. Remembering that in the present context the
chi-squared test is naturally two-sided, the evidence against CSR is weak;
further support for this conclusion is provided by the fact that for both 4×4
and 2×2 grids the observed value of X^2 is close to its expectation under CSR.

2.5.2 Analysis of redwood seedlings

For the 62 redwood seedlings a 3×3 grid is again a reasonable choice. The
observed counts are

$$
\begin{matrix}
5 & 9 & 6 \\
13 & 8 & 2 \\
0 & 6 & 13
\end{matrix}
$$

and the X^2 value of 22.77 is highly significant ($p = 0.0037$). A 4×4 grid
similarly leads to emphatic rejection of CSR ($p = 0.0010$), whereas a 2×2
grid gives $p = 0.156$. A plausible model for these data, which we investigate
further in Chapter 7, involves randomly distributed clusters of events. In these
circumstances it is not unreasonable that the smaller-sized quadrats give the
stronger rejection of CSR. The essential point to note here is that the choice
of quadrat size can have a marked effect on the result of the test.

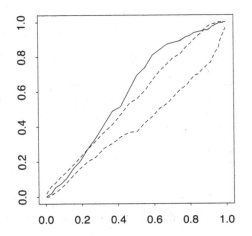

FIGURE 2.10

EDF plot of point to nearest event distances for biological cells. ——— : data; − − − : upper and lower envelopes from 99 simulations of CSR.

2.5.3   Analysis of biological cells

For the 42 biological cells, the observed values of X^2 are below expectation for 2×2, 3×3 and 4×4 grids, significantly so in the 2×2 and 4×4 cases, although for a 4×4 grid the expected frequencies under CSR are a little small for comfort. The failure to reject CSR using the 3×3 grid suggests that the test is weak against regular alternatives to CSR. For comparison with the previous two examples, we give the 3×3 array of counts,

$$
\begin{array}{ccc}
3 & 6 & 3 \\
4 & 7 & 6 \\
3 & 6 & 4 \\
\end{array}
$$

2.6   Scales of pattern

Greig-Smith (1952) proposed the following method for the analysis of data presented as counts in a large grid of contiguous quadrats. The sample variance-to-mean ratio, also called the index of dispersion, of the counts is first calculated for the basic grid and for further grids obtained by successive combination of adjacent quadrats into 2×2, 4×4, etc., *blocks*. The index of dispersion is then plotted against block size and peaks or troughs in the graph are interpreted as evidence of *scales of pattern* (aggregated or regular, respectively).

This method of analysis originated in plant ecology, in which field it became extremely popular; a review by Greig-Smith (1979) lists numerous applications. A possible objection to the method is that formal tests for the significance of peaks and troughs in the sequence of indices of dispersion at different scales are not available. According to the development in Section 2.5, each index is proportional to a chi-squared statistic for testing the hypothesis that the events are an independent random sample from the uniform distribution over the study region. Either this hypothesis is true or it is false – it makes no sense to ask whether it is true at some scales and false at others.

Mead (1974) addressed this formal defect of the Greig-Smith procedure by establishing a series of independent tests for pattern at several scales. Mead's procedure requires the data to be partitioned successively into 1, 4, 16, etc., blocks each consisting of 16 counts in a 4×4 grid. At each stage, the hypothesis to be tested is that, within each block, the set of counts in the four associated 2×2 sub-blocks is a random selection from $(16!)/(4!)^5 = 2,627,265$ equally likely possibilities, as implied by CSR. Mead's suggested test statistic is the sum of the six absolute pairwise differences between the four sub-block counts within a block, summed in turn over all blocks. A significantly large value of this statistic implies that, within blocks, counts in neighbouring quadrats are relatively similar, and this would be interpreted as evidence of aggregation at the appropriate scale. A significantly small value similarly implies relatively dissimilar counts in neighbouring quadrats. This is more difficult to interpret. An extreme manifestation of it would be a chess-board pattern of alternating high and low counts, but this seems unlikely to arise in practice. Once the test statistic, u say, has been chosen, the test itself can be implemented via Monte Carlo sampling of the null randomization distribution of u. The independence of the tests at the various scales follows because the randomized 2×2 sub-block counts at one scale become the fixed 4×4 block counts at the next smaller scale, and so on.

So far, we have assumed that the quadrat grid has seen superimposed retrospectively on a mapped pattern. In practice, the counts may be recorded directly in the field and a common variation, proposed by Kershaw (1957), is to replace the $k \times k$ grid by an $m \times 1$ transect. The data are then being analysed essentially as a time-series, an analogy which is strengthened by Ripley's (1978) interpretation of Greig-Smith's method as a form of spectral analysis on the quadrat counts. By applying Greig-Smith's, Mead's and related methods to simulated data, Ripley also shows that the results can be difficult to interpret in terms of an underlying generating mechanism. From a modern statistical perspective, this is of greater concern than the lack of formal tests of significance.

2.6.1 Analysis of Lansing Woods data

Gerrard (1969) describes an investigation of a 19.6 acre square plot in Lansing Woods, Clinton County, Michigan, USA. In particular, he has provided the

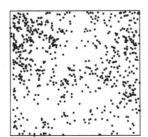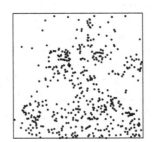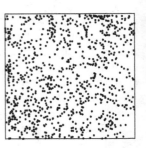

FIGURE 2.11
Locations of trees in Lansing woods: left panel, hickories; middle panel, maples; right panel, oaks.

locations of 2251 trees in the plot. Maps of these data are shown, for the three major species groupings of hickories, maples and oaks, in Figure 2.11. Each map is here converted to counts in a 32 × 32 grid, which permits the investigation of four scales of pattern using Mead's procedure. The tests are implemented with 99 Monte Carlo randomizations, and the test statistic at each scale is the one suggested by Mead.

Table 2.1 gives the ranks of the observed test statistics amongst the Monte Carlo randomizations. For each of the three species groupings there is moderate or strong evidence of aggregation at the smallest scale and a strong indication of aggregation at one or more further scales. In the cases of the hickories and maples, departure from CSR is obvious from inspection of the data, but for the oaks the visual impression is less clear. Figure 2.12 shows a plot of the quadrat count index of dispersion against block size, for each of the three species groupings. The sequence of generally increasing values of the index observed in each case is consistent with an underlying mechanism involving random variation in the local intensity of events, and in Chapters 7 and 8 we shall re-examine the data from this viewpoint. The plots in Figure 2.12 do not relate in any obvious way to the results of the analysis by Mead's procedure. This inconsistency was experienced also by Ripley (1978). It cannot be dismissed as an artefact of the Monte Carlo testing, but rather implies that the concept of a scale of pattern is somewhat ill-defined.

2.6.2 Scales of dependence

Besag (1978) describes a bivariate analogue of Mead's test in order to investigate what we might term *scales of dependence* between two patterns. For this, the basic unit is a 2 × 2 block of quadrats. Each quadrat provides a pair of counts, one for each species. Besag suggests testing the hypothesis that the

TABLE 2.1

Analysis of Lansing Woods data by Mead's procedure, using a 32×32 grid of quadrat counts and 99 randomizations. A high rank for u_1 suggests aggregation. Where ties occur, the least extreme rank is quoted. Block size refers to the number of quadrats which are treated as a single 4×4 block.

Rank of u_1 amongst	Block size			
$u_i : i = 1, ..., 100$	4×4	8×8	16×16	32×32
Hickories	99	100	100	90
Maples	100	71	100	62
Oaks	95	99	90	73

two sets of four counts within a block are independent, using as test statistic the Spearman rank correlation coefficient (Kendall, 1970; Sprent, 1981, Chapter 10) for the two sets of four, summed over all blocks. The test is implemented via Monte Carlo randomization of counts within blocks. Each 2×2 block is then aggregated into a single quadrat, and so on, to give a sequence of independent tests of the hypothesis of independence for the two sets of counts within blocks. Besag applies this procedure to the Lansing Woods data and detects negative dependence between the hickories and maples, at the single scale corresponding to a partitioning of the study region into a 4×4 grid of blocks, i.e. an 8×8 grid of quadrats. The analysis of patterns formed by two or more distinct types of event will be considered in more detail in later chapters.

As noted in Section 1.2, contiguous quadrat counts represent a form of data intermediate between complete mapping and the sparse sampling procedures that we shall discuss in Chapter 3. Greig-Smith (1979) and Mead (1974) have emphasized that their analyses are intended to be exploratory in nature. As such, they are most useful for large, potentially heterogeneous data-sets, particularly if the results suggest further, and quite possibly non-statistical, investigations of the underlying processes.

2.7 Recommendations

The discussion in this chapter falls short of any systematic investigation of the power of the various tests under consideration. During the early development of the subject, many publications focused on tests of CSR and a number included comparative power assessments; see, for example, Ripley and Silverman (1978), Diggle (1979a) and Ripley (1979a). Our recommendations are based

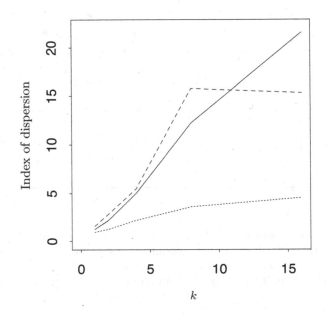

FIGURE 2.12
Index of dispersion plotted against block size $(k \times k)$ for Lansing Woods data, based on a 32×32 grid of quadrat counts. —— : hickories; – – – : maples; .... : oaks.

partly on a synthesis of published results, but also on accumulated practical experience.

Our over-riding view is that a test of CSR is of very limited inherent interest, but rather should be seen as a framework within which exploratory analysis can be conducted. In our view, the most useful procedures are those based on functional summary descriptions of the data together with simulation envelopes to indicate the range of statistical variation under CSR. Of the three functional summaries considered in this chapter, we recommend using both $\hat{F}(\cdot)$ and $\hat{G}(\cdot)$ routinely. The two corresponding theoretical distribution functions, $F(\cdot)$ and $G(\cdot)$, are equal if the underlying point process is a homogeneous Poisson process, i.e. if CSR prevails. Also, departures from CSR typically induce opposite deviations in $\hat{F}(\cdot)$ and $\hat{G}(\cdot)$ from their common theoretical form under CSR. These deviations show up in the main body of each distribution, and are therefore easily seen from their graphical representations as EDF plots. For these reasons, it may be useful to combine the two types of nearest neighbour distance into a single test statistic. A possible test statistic, analogous to those used in Sections 2.3 and 2.4, would be

$$u_1 = \int \{\hat{F}(x) - \hat{G}(x)\}^2 dx$$

Another, suggested by Van Lieshout and Baddeley (1996), would be a statistic based on an estimate of the function $J(x) = \{1 - G(x)\}/\{1 - F(x)\}$. In this case, it is not necessary to use edge-corrected estimators for $F(\cdot)$ and $G(\cdot)$, as Van Lieshout and Baddeley have shown that the estimate of $J(\cdot)$ is insensitive to edge-effects.

The functions $\hat{F}(\cdot)$ and $\hat{G}(\cdot)$, either separately or in combination, are also useful for assessing the goodness-of-fit of a date-set to any stochastic model. One reason for this is that they are complementary to the second-order methods and likelihood-based methods that, as we shall see in later chapters, are more useful for identifying a suitable class of models and estimating its parameters, once CSR has been rejected.

The inter-event distance distribution function, $H(\cdot)$, is less useful for preliminary testing. Its behaviour in our three illustrative examples, in which only the lower tail of the distribution is sensitive to quite pronounced changes in the underlying pattern, is typical. However, the distribution of inter-event distances is closely related to the second-order properties of a spatial point process and, as we shall see in later chapters, in this form it is a valuable tool in the much wider context of formulating and fitting stochastic models.

Quadrat count methods are now used infrequently for the analysis of mapped data. By comparison with the distance-based methods, they are less easily adaptable to more ambitious tasks, such as parameter estimation within a declared class of stochastic models.

3

Methods for sparsely sampled patterns

CONTENTS

3.1 General remarks

In this chapter we consider methods for the analysis of data obtained by a sparse sampling procedure, as defined in Section 1.2. We recall that such data consist either of *quadrat counts* in small areas within a study region A, or of distances measured from sampling points in A to neighbouring events. The number n of events in A is unknown, but must be assumed to be very much larger than m, the number of quadrats or sampling points, otherwise a complete mapping of A would presumably be feasible and would certainly be more informative. Typically, A will be large and potentially heterogeneous. For example, sparse sampling methods were originally devised as field-sampling methods in forestry and plant ecology.

The objectives of a sparse sampling analysis will usually be to estimate the number of events in A, or equivalently the *intensity*, defined to be the mean number of events per unit area, and to obtain a qualitative description of the underlying pattern through the application of one or more tests of complete spatial randomness. If more detailed inferences are required, these are better dealt with by the collection and analysis of mapped data in sub-regions of

A. Indeed, one useful function of a preliminary, sparse sampling analysis is to provide guidelines for the planning of a subsequent more detailed investigation.

Most theoretical discussions of sparse sampling assume, if only implicitly, that sample points are randomly located according to a uniform distribution on A. However, the essential practical requirement for the validity of most of the associated statistical methods that we consider in this chapter is that sample points should be well separated in order that observations from different points can be assumed to be independent. This is most easily achieved by a systematic sampling design. Another desirable feature of systematic sampling is that it conveniently allows for retrospective partitioning of A into several sub-regions within which separate analyses can be performed and the results compared. A possible disadvantage is that the interval between successive sample points may coincide with periodicities in the underlying pattern. If this is thought to be a serious danger, it can be alleviated by random sampling within sub-regions of A. Other possible designs include, for example, the location of sample points along line transects. Note that the statistical merits of different sampling schemes should be compared on the basis of equal effort in the field. In this respect, systematic sampling may permit a larger value of m and, all other things being equal, a more sensitive analysis. In theory, the maximum value of m is constrained by the requirement that different sample points should generate independent observations. In practice, this constraint is not severe if systematic sampling is used. Byth and Ripley (1980) use simulations to establish that for any of the commonly used distance measurements, m may be at least as large as $0.1n$. If random sampling is used a safe upper limit is about $0.05n$. In the remainder of this chapter we assume without further comment that observations from different sample points are independent. Whatever sampling design is adopted, objectivity in the positioning of the sample points is, of course, vital.

In the remainder of this chapter, we first discuss separately the use of quadrat counts and distance methods for testing CSR and for estimating intensity. We then describe tests of independence between pairs of patterns, which are relevant to the interpretation of data like the Lansing Woods data in which different species are recorded within the same study region.

3.2   Quadrat counts

We recall from Section 1.3 that, under CSR, the number $N(B)$ of events in any region with area B follows a Poisson distribution with mean λB, where λ is the intensity. Explicitly, the probability distribution of $N(B)$ is

$$p_n(B) = \exp(-\lambda B)\{(\lambda B)^n/n!\} : n = 0, 1, 2, \ldots \tag{3.1}$$

In this section, we assume that the available data comprise independent counts $n_1, n_2, ..., n_m$ in m such quadrats, each of area B.

3.2.1 Tests of CSR

We wish to test the hypothesis that the n_i are an independent random sample from a Poisson distribution with unspecified mean. A natural test statistic is the sample variance-to-mean ratio or *index of dispersion*,

$$I = \sum_{i=1}^{m} (n_i - \bar{n})^2 / \{(m-1)\bar{n}\}. \tag{3.2}$$

The intuitive appeal of (3.2) rests on the equality of the mean and variance of the Poisson distribution (3.1). Thus, I can be interpreted as a variance ratio statistic. The numerator,

$$s^2 = (m-1)^{-1} \sum_{i=1}^{m} (n_i - \bar{n})^2,$$

estimates the variance of $N(B)$ when no distributional assumptions are made, whilst the denominator, $\bar{n}$, estimates the variance when CSR holds. The index of dispersion was first used by Fisher *et al.* (1922). Under CSR, the sampling distribution of $(m-1)I$ is χ^2_{m-1}, to a good approximation provided that $m > 6$ and $\lambda B > 1$ (Kathirgamatamby, 1953). Significantly large or small values respectively indicate aggregated or regular departures from CSR. In Section 2.5 we showed that $(m-1)I$ could also be interpreted as Pearson's goodness-of-fit criterion for a uniform distribution of events over the union of the m quadrats, conditional on the total count.

The power of the index of dispersion test obviously increases with m, but also depends in an unpredictable way on the size and shape of the individual quadrats. Perry and Mead (1979) calculate the power of the test against a heterogeneous alternative to CSR involving interspersed patches of high and low intensity. Stiteler and Patil (1971) calculate the theoretical variance-to-mean ratio for some regular lattice patterns. Results in the above two papers and in Diggle (1979a) suggest that the index of dispersion test is generally powerful against aggregated alternatives to CSR, but may be weak against regularity.

For quadrat count data, the index of dispersion appears to have no serious rivals as a test statistic. Cormack (1979) notes that alternative indices proposed by Morisita (1959) and by Lloyd (1967) need to be converted to $(m-1)I$ in order to test CSR.

Early work on the analysis of quadrat count data concentrated on the development of more general families of discrete distributions than the single-parameter Poisson, especially with a view to modelling aggregated patterns. See, for example, Evans (1953) or Douglas (1979). The story of the rise and fall of these so-called "contagious distributions" as tools for the analysis of spatial data is of some historical interest because it shows how attempts to model observed data through discrete *distributions* ultimately foundered on their failure to respect the underlying setting of spatial point *processes*. We shall return to this in Chapter 6, as part of a wider discussion of the various classes of models that have been proposed for aggregated patterns.

3.2.2   Estimators of intensity

An intuitively reasonable estimator of the intensity is the total count divided by the total quadrat area,

$$\hat{\lambda} = \sum_{i=1}^{m} n_i/(mB). \tag{3.3}$$

It is easy to establish from (3.1) that this is the maximum likelihood estimator under CSR, in which case λ is unbiased for $\hat{\lambda}$, with variance $\lambda/(mB)$. More generally, $\hat{\lambda}$ is always an unbiased estimator for the intensity but its variance may depend on the size and shape of the individual quadrats and on the sampling scheme, as well as on the total quadrat area. Note that, strictly, the variance of $\hat{\lambda}$ is different according to whether it is regarded as an estimator for the observed number of events per unit area within A or for the intensity of an underlying spatial point process that is assumed to have generated the observed pattern: in the former case the variance goes to zero as a systematic sample of quadrats extends to cover the whole of A. In practice, the assumed sparseness of the sampling makes this distinction unimportant and the sample standard deviation of the observed counts can be used to construct interval estimates of λ. Some authors, including Ghent (1963), have suggested that in practice $\hat{\lambda}$ may be biased by a tendency for the field-worker to include events just outside the individual quadrat boundaries, in the mistaken belief that an empty quadrat contains no information.

3.2.3   Analysis of Lansing Woods data

We now apply these techniques to the Lansing Woods data introduced in Section 2.7, taking a systematic sample of 25 square quadrats of side 0.05 arranged in a 5×5 grid. We emphasise that this analysis is purely illustrative, because analysing a mapped pattern by sparse sampling methods is inefficient. Table 3.1 gives the results for the analysis of each of the three species groupings. For the hickories and maples, CSR is overwhelmingly rejected in favour of an aggregated alternative, whilst for the oaks CSR is accepted (the one-sided 5% and 1% critical values of I when $m = 25$ are 1.52 and 1.79, respectively). In all three cases, $\hat{\lambda}$ is within one empirical standard error of λ, which we take to be equal to n because A is the unit square.

In these analyses, the individual counts are typically small (average counts

TABLE 3.1
Quadrat count analysis of Lansing Woods data, using 25 square quadrats of side 0.05.

	λ	I	$\hat{\lambda}/\lambda$	$S.E.(\hat{\lambda}/\lambda)$
Hickories	703	2.59	1.02	0.25
Maples	514	2.92	0.90	0.29
Oaks	929	1.22	1.03	0.15

for hickories, maples and oaks were 1.80, 1.16 and 2.40 respectively) but the quadrats are nevertheless physically quite large, as 0.05 translates to about 46 feet (14 metres) in the field. We repeated the analysis using 100 quadrats of side 0.02 and found marginal evidence against CSR for the maples and the oaks, but none at all for the hickories. However, the null distribution theory is suspect in this case because of the small average counts.

These results are generally consistent with, but weaker than, those obtained in Section 2.6 using a grid of contiguous quadrats to partition the whole of the study region.

3.3 Distance measurements

Distance methods, also known as plot-less sampling techniques, were introduced because of the practical difficulties sometimes raised by quadrat sampling. An early reference is Cottam and Curtis (1949). It is straightforward to devise a large number of subtly different distance methods, each with its own distribution theory. Possibly for this reason, an extensive literature on distance methods developed throughout the 1950's, 1960's and 1970's. Most of the early work was concerned with the definition of various types of distance measurement and associated statistics to test CSR or to estimate intensity. Holgate (1965a) marked something of a departure in that he evaluated the power functions of several tests of CSR against theoretical alternatives, thus providing an objective basis for the choice of a method. Developments since 1965 tended to continue in this vein, investigating the power of tests of CSR (Holgate, 1965b; Besag and Gleaves, 1973; Brown and Holgate, 1974; Diggle *et al.*, 1976; Cox and Lewis, 1976; Diggle, 1977b; Hines and Hines, 1979; Byth and Ripley, 1980) or the robustness of estimators of intensity (Persson, 1971; Pollard, 1971; Holgate, 1972; Diggle, 1975, 1977a; Cox, 1976; Warren and Batcheler, 1979; Patil *et al.*, 1979; Byth, 1982).

3.3.1   Distribution theory under CSR

When CSR holds, the distribution theory for the various distance methods can be derived from the Poisson distribution of quadrat counts together with the independence of counts in disjoint regions. From (3.1), taking B to be the area πx^2 of a disc of radius x, we immediately deduce that the distribution function of the distance X from an arbitrary point (or event) to the nearest (other) event is

$$F(x) = 1 - \exp(-\pi\lambda x^2) : x \geq 0, \tag{3.4}$$

a result previously given at (2.8) and (2.12). Notice that πX^2 follows an exponential distribution with parameter λ and $2\pi\lambda X^2$ is therefore distributed as χ_2^2.

Various other distance distributions associated with the Poisson process can be derived from (3.1) and the independence of numbers of events in disjoint regions. Let $X_{k,\theta}$ denote the distance from an arbitrary point or event to the kth nearest event within a sector of included angle $\theta \leq 2\pi$ and arbitrary orientation. Let $U_k = \frac{1}{2}\theta X_{k,\theta}^2$ and note that U_k is the area of a sector of included angle θ and radius $X_{k,\theta}$. Then,

$$P(U_1 > u_1) = P\{N(u_1) = 0\} = e^{-\lambda u_1},$$

again using (3.1). Furthermore, for any $u_2 > u_1$,

$$P(U_2 > u_2 \mid U_1 = u_1) = P\{N(u_2 - u_1) = 0\} = \exp\{-\lambda(u_2 - u_1)\},$$

so that the conditional probability distribution function (pdf) of U_2, given $U_1 = u_i$, is $\lambda \exp(-\lambda(u_2 - u_1))\}$ and the joint pdf of (U_1, U_2) is

$$f_2(u_1, u_2) = \lambda^2 \exp(-\lambda u_2) : 0 < u_1 < u_2.$$

Essentially the same argument gives the joint pdf of $(U_1, ..., U_k)$ for any k as

$$f_k(u_1, ..., u_k) = \lambda^k \exp(-\lambda u_k) : 0 < u_1 < \cdots < u_k, \tag{3.5}$$

a result due to Thompson (1956).

More intricate geometrical constructions can be handled similarly. For example, Cox and Lewis (1976) consider the joint distribution of random variables X and Y defined as the respective distances from an arbitrary point O to the nearest event, at P say, and from P to the nearest other event, Q. Then,

$$P(Y > y \mid X = x) = P[N\{A(x, y)\} = 0],$$

where

$$A(x, y) = \pi y^2 - (\phi y^2 + \theta x^2 - xy \sin \phi) \tag{3.6}$$

is the area of the shaded region in Figure 3.1, $\cos \phi = y/(2x)$ and $\theta + 2\phi = \pi$.

Notice in particular that P is *not* an arbitrary event; the selection procedure for P is biased in favour of the more isolated events in the population.

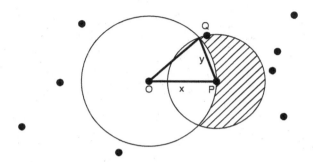

FIGURE 3.1
The nearest event, P, to an arbitrary point, O, and the nearest event, Q, to
the event P.

Cox and Lewis further deduce from (3.6) that, conditional on $Y > 2X$, the
random variable $4X^2/Y^2$ is uniformly distributed on $(0, 1)$. Cormack (1977)
uses a strikingly simple geometrical argument to show that this holds for *any*
underlying pattern. Briefly, for an arbitrary pattern of events, consider the
ith event to be located at the centre of a disc of radius $x_i/2$, where x_i is the
distance from the ith event to its nearest neighbour. These circles touch when
pairs of events are reciprocal nearest neighbours, but cannot intersect. Conditioning on $Y > 2X$ is equivalent to placing the sampling origin O uniformly
at random within the union of these discs, and the result follows.

A related, but distributionally simpler, device is the T-square sampling
procedure of Besag and Gleaves (1973). As shown in Figure 3.2, O and P are
as above, but Q is now the nearest event to P under the restriction that the
angle OPQ must be at least $\pi/2$. With $X = $OP as above and $Z = $PQ, we see
that

$$P(Z > z | X = x) = P\{N(\pi z^2/2) = 0\} = \exp(-\lambda \pi x^2/2)$$

and deduce that $2\pi\lambda X^2$ and $\pi\lambda Z^2$ are independently and identically distributed as χ_2^2

T-square and other distance-based methods for sampling spatial point patterns were originally applied in forestry and plant ecology, but have recently
re-emerged as a tool for surveying refugee camps and other unplanned human
settlements (Bostoen, Chalabi and Grais, 2007; see also Section 3.3.5).

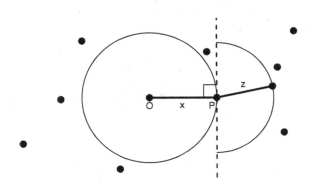

FIGURE 3.2
T-square sampling: the nearest event, P, to an arbitrary point, O, and the T-square nearest neighbour, Q, to the event P.

3.3.2   Tests of CSR

In this sub-section, remarks on the comparative power of different tests represent an overview of results in Diggle *et al.* (1976), Hines and Hines (1979) and Byth and Ripley (1980). The original papers give more details in specific instances.

One general approach to the construction of a scale-free statistic to test CSR is to compare two types of distance measurement. For example, Hopkins (1954) considers measurements $x_i : i = 1, ..., m$, from each of m sample points to the nearest event and $y_i : i = 1, ..., m$ from each of m randomly sampled events to the nearest other event. Under CSR and sparse sampling, $2\pi\lambda x_i^2$ and $2\pi\lambda y_i^2$ are independently distributed as χ_2^2. Thus, $2\pi\lambda \sum_{i=1}^{m} x_i^2$ and $2\pi\lambda \sum_{i=1}^{m} y_i^2$ are independently distributed as χ_{2m}^2 and

$$h = \sum x_i^2 / \sum y_i^2 \qquad (3.7)$$

is distributed as $F_{2m,2m}$. In (3.7), and in the remainder of this section unless stated otherwise, summations are over $i = 1, ..., m$. The rationale for Hopkins' test is that in an aggregated pattern, the point-event distances x_i will be large relative to the event-event distances y_i, and *vice versa* in a regular pattern. Thus, significantly large or small values of h indicate aggregation or regularity, respectively. Note that $h/(1 + h)$ lies between 0 and 1 and could therefore be regarded as an index of pattern. The statistic h was proposed independently by Moore (1954).

Within the context of sparse sampling methods, Hopkins' test generally has good power properties. However, the random selection of events requires a complete enumeration within A, which is precisely the operation we wish to avoid in a sparse sampling analysis. Byth and Ripley (1980) propose implementing Hopkins' test, but selecting one event at random from each of m quadrats "of a size which would contain about five trees on average". In some contexts this might still prove impractical, and would in any case imply an increased effort in the field which has not been allowed for in published power comparisons.

Holgate (1965b) considers measurements $(x_{1i}, x_{2i}) : i = 1, ..., m$ from each of m sample points to the nearest and second nearest events, respectively. He then uses (3.5) to deduce that, under CSR, x_{1i}^2/x_{2i}^2 is uniformly distributed on $(0, 1)$, whence

$$h_N = m^{-1} \sum (x_{1i}^2/x_{2i}^2)$$

is Normally distributed with mean 1 and variance $(12m)^{-1}$, to an excellent approximation if $m > 10$. Alternatively,

$$h_F = \sum x_{1i}^2 / \sum (x_{2i}^2 - x_{1i}^2)$$

is distributed as $F_{2m,2m}$. The rationale for h_F is that $(x_{2i}^2 - x_{1i}^2)$ should behave like y_i^2, and the interpretation of h_N or h_F is the same as for Hopkins' h. Published results suggest that Holgate's tests are generally less powerful than Hopkins' test and, in particular, are very weak against regular alternatives to CSR.

Eberhardt (1967) considered only point–event distances x_i, and proposed an index

$$e = m \sum x_i^2 / (\sum x_i)^2.$$

Note that $\sqrt{\{m(e-1)/(m-1)\}}$ is the sample coefficient of variation of the distances x_i. Hines and Hines (1979) provide critical values to test CSR but the test must be weak against aggregated alternatives, because the distribution of e under CSR applies also to a process of randomly distributed point clusters, in which each single event of a completely random pattern is replaced by a fixed or random number of coincident events. By the same argument, any scale-free statistic based only on measurements of the distance from a sample point to the nearest event must be suspect.

Besag and Gleaves (1973) use data $(x_i, z_i) : i = 1, ..., m$ generated by their T-square sampling procedure as described in Section 3.3.1. Recall that x_i is a point-event distance and z_i an event-event distance in a restricted area of search. Two possible test statistics are

$$t_N = m^{-1} \sum x_i^2 / (x_i^2 + z_i^2/2) \tag{3.8}$$

and

$$t_F = 2 \sum x_i^2 / \sum z_i^2, \tag{3.9}$$

whose distributions under CSR are the same as for the corresponding Holgate statistics h_N and h_F. Significantly large or small values again suggest aggregation or regularity, respectively. The T-square tests are generally less powerful than Hopkins' test but more powerful than their Holgate counterparts, particularly against regular alternatives. On balance, t_N is preferable to t_F, although this does depend on the range of alternatives under consideration. Hines and Hines (1979) recommend a variant of Eberhardt's index based on T-square sampling. In this variant, the measurements x_i and $z_i/\sqrt{2}$, which under CSR are independent with a common distribution given by (3.4), are treated as a single sample of size $2m$. The resulting test statistic appears to be slightly more powerful than t_N against a range of aggregated and regular alternatives, although its interpretation is less transparent.

One advantage of the T-square sampling procedure is that the simplicity of its distribution theory under CSR allows to some extent for the development of appropriate tests when the range of alternatives is restricted a priori. For example, a test based on t_N is insensitive to long-range fluctuations in local intensity. Specifically, if CSR applies locally but with possibly different values, $\lambda_i : i = 1, ..., m$ say, of the intensity parameter associated with the different sample points, the distribution of t_N is the same as under CSR. However, within this restricted context it is straightforward to derive the likelihood ratio test of CSR, which corresponds to equal λ_i. The test statistic is

$$M = 48m \left\{ m \log(\bar{u}) - \sum \log u_i \right\} / (13m + 1) \qquad (3.10)$$

where $u_i = x_i^2 + z_i^2/2$. The approximate distribution of M under CSR is χ^2_{m-1}. This test is a direct analogue of Bartlett's (1937) test of the equality of variances in Normal sampling and incorporates the correction factor recommended by Bartlett to improve the chi-squared approximation. Note that the test is one-sided: significantly large values suggest rejection of CSR. Diggle (1977b) proposed a two-stage procedure in which the M-test is applied only if an initial test using t_N gives a non-significant result. The effect of this is to achieve a four-way classification of the underlying pattern as regular, random, heterogeneous or aggregated, although the two-stage procedure means that the nominal significance level for M is not strictly correct.

Another situation in which the standard tests are unsatisfactory is when elements of aggregation and regularity are combined. For example, Brown and Rothery (1978) discuss the detection of local regularity in the presence of long-range aggregation, motivated by a study of the spacing of ducks' nests when only a small proportion of the study region contains usable nesting sites (Brown, 1975). In this context, T-square sampling will typically generate point-event distances x_i which are larger than the event-event distances z_i, indicating aggregation. The local regularity could be detected by using only the z_i. Two possible statistics based on the $z_i : i = 1, ..., m$ are the Eberhardt index or a variant of the M statistic,

$$M = 24m \left\{ m \log(\bar{z^2}) - \sum \log z_i^2 \right\} / (7m + 1)$$

with significantly *small* values of either statistic indicating regularity.

Cox and Lewis (1976) work with data $(x_i, y_i) : i = 1, ..., m$, where x_i is the distance from the ith sample point to the nearest event and y_i the distance from that event to the nearest other event. As described in Section 3.3.1 above, Cormack (1977) shows that pairs (x_i, y_i) with $y_i > 2x_i$ are uninformative. Cox and Lewis consider the $m_0 \le m$ pairs for which $y_i < 2x_i$, and use (3.6) to show that the following sequence of transformations produces observations $r_i : i = 1, ..., m$, whose distribution under CSR is uniform on $(0, 1)$:

(i) $\theta_i = 2 \sin^{-1}\{y_i/(2x_i)\}$;

(ii) $w_i = \{2\pi + \sin\theta_i - (\pi + \theta_i)\cos\theta_i\}^{-1}$;

(iii) $r_i = (4\pi w_i - 1)/3$. It follows that $cl = m_0^{-1}\sum r_i$ is approximately Normally distributed with mean 0.5 and variance $(12m_0)^{-1}$ under CSR. Significantly large or small values indicate aggregation or regularity respectively, and in either case the power of the test appears to be comparable to that of t_N.

3.3.3 Estimators of intensity

Suppose now that distances are measured from each sample point to the nearest, second nearest, ..., kth nearest event. Then, (3.5) shows that under CSR the distances $x_{ki} : i = 1, ..., m$ to kth nearest events are sufficient for λ, and the maximum likelihood estimator of $\gamma = \lambda^{-1}$ is

$$\hat{\gamma}_k = \pi(\sum x_{ki}^2)/(km)$$

which is unbiased with variance $\gamma^2/(km)$. An increase in the value of k gives an estimator which has smaller variance, but whose application in the field is more time-consuming. The more subtle question of robustness to departures from CSR will be considered shortly. The change from λ to γ as the parameter of interest makes for ease of presentation, but also seems natural for a distance-based method of estimation, since squared distances effectively measure areas. Holgate (1964) showed that if the total quadrat area in the quadrat count estimator (3.5) is set equal to the expected area of search involved in computing $\hat{\gamma}_k^{-1}$, considered as an estimator for λ, then the two methods are equally efficient under CSR and, incidentally, all choices of k in $\hat{\gamma}_k$ are equally efficient.

The major objection to $\hat{\gamma}_k$ is that it can be seriously biased when CSR does not hold (Persson, 1971; Pollard, 1971). The bias tends to become smaller for larger values of k, but identifying the kth nearest neighbour in the field then becomes difficult. An alternative strategy for reducing the bias is to note that estimators based on point-event and on event-event distances tend to be biased in opposite directions. An average of the two should therefore be more robust, in the sense of having smaller bias for a wide range of underlying patterns. Under CSR, the maximum likelihood estimator for γ based on a T-square sample $(x_i, z_i) : i = 1, ..., m$, is

$$\hat{\gamma}_T = \pi(\sum x_i^2 + \sum z_i^2/2)/(2m),$$

the arithmetic mean of estimators based on the x_i or z_i measurements separately. Results in Diggle (1975, 1977a) suggest that a more robust estimator is

$$\gamma_T^* = (\pi/m)\sqrt{(2\sum x_i^2/\sum z_i^2)}$$

whilst Byth (1982) recommends

$$\tilde{\gamma}_T = (2\sqrt{2}/m^2)(\sum x_i \sum z_i), \tag{3.11}$$

which is less sensitive than γ_T^* to an occasional very large x_i measurement in a strongly aggregated pattern.

Because $\tilde{\gamma}_T$ is a function of two sample means, its approximate standard error can be calculated using the delta technique. Let $\bar{x}$, $\bar{z}$, s_x^2, s_z^2 and s_{xz} denote sample means, variances and covariance. Then,

$$S.E.(\tilde{\gamma}_T) \approx \sqrt{\{8(\bar{z}^2 s_x^2 + 2\bar{x}\bar{z}s_{xz} + \bar{x}^2 s_z^2)/m\}}.$$

Strictly, this does not lead to interval estimates for γ but for $E(\tilde{\gamma}_T)$. However, it would appear that the bias of $\tilde{\gamma}_T$ is often small. It is admittedly easy to devise theoretical point process models for which $\tilde{\gamma}_T$ performs badly, but these typically represent extreme departures from CSR which would be easily detected in the field. One example would be a process involving large, tightly clustered groups of events, in which case it would be more sensible to estimate separately the mean area per cluster and the mean cluster size.

Cox (1976) and Warren and Batcheler (1979) adopt a different approach in which an empirically determined correction factor is applied to an estimator based only on point-event distances. Warren and Batcheler report a variety of successful applications, but Byth (1982) obtains disappointing results from a simulation study.

Patil *et al.* (1979) devise a consistent estimator for the intensity of any stationary process that does not generate multiple coincident events, but this theoretically desirable property appears to have been achieved at the expense of a large increase in variance.

3.3.4   Analysis of Lansing Woods data

We again use the Lansing Woods data to illustrate the use of sparse sampling techniques, in this case T-square sampling. We take a systematic sample of $m = 25$ points in a 5×5 grid. To test CSR we use the t_N-statistic (3.8) followed if necessary by the M statistic (3.10). To estimate γ, the mean area per event, we use Byth's estimator $\tilde{\gamma}_T$, defined at (3.11). The results are given in Table 3.2. We accept CSR for the oaks, but reject CSR in favour of a heterogeneous

TABLE 3.2

T-square analysis of Lansing Woods data. Figures in parentheses indicate attained significance levels for tests of CSR (two-sided for t_N, one-sided for M).

	γ	t_N	M	$\tilde{\gamma}_T/\gamma$	$S.E.(\tilde{\gamma}_T/\gamma$
Hickories	0.001425	0.54 (0.53)	41.26 (0.02)	1.47	0.26
Maples	0.001946	0.60 (0.08)	38.28 (0.03)	1.23	0.29
Oaks	0.001076	0.42 (0.17)	32.18 (0.12)	0.99	0.15

alternative for both the hickories and the maples. The estimates of $\tilde{\gamma}_T$ are within one empirical standard error of γ except for the hickories, where the difference is about 1.8 standard errors.

3.3.5 Catana's wandering quarter

An ingenious sampling procedure whose statistical potential appears not to have been tapped is Catana's (1963) "wandering quarter." This generates a single point-event distance x_0 and a sequence of event-event distances x_i : $i = 1, ..., m$, from a single starting point O, as indicated in Figure 3.3. Under CSR, the transformed observations πx_i^2 are independently distributed as χ_2^2. Inferential procedures could, and should, recognize the spatial ordering of the x_i. If the underlying pattern is one of patchy spatial heterogeneity, this would induce autocorrelation in the series. Even without any formal analysis, simple plotting of the x_i in sequence order could reveal interesting spatial trends.

The method could also be used to sample a large region by a series of essentially parallel traverses. From a practical point of view, it is attractive that the field-worker can take observations continuously as they traverse the study region, whilst the restriction of the area of search for each event to a ninety-degree sector neatly ensures that successive regions of search do not overlap.

In their discussion of distance-based sampling methods for estimating the size of unplanned human settlements such as refugee camps, Bostoen, Chalabi and Grais (2007) mention Catana's method but focus on T-square sampling, including a detailed discussion of how to optimise a route through a series of sampling origins distributed at random within the region of interest. In this context Catana's method, which generates a whole sequence of measurements from a single sampling origin, would seem to have considerable practical advantages over T-square sampling in terms of the amount of information that can be gathered for a fixed sampling effort in the field.

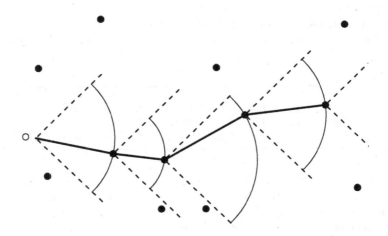

FIGURE 3.3
Catana's wandering quarter: solid lines indicate a sequence of recorded distances, starting from the point O.

3.4   Tests of independence

When events of two different types, for example plants of different species, co-exist within a study region, it may be of interest to establish whether the two underlying point processes are independent.

For quadrat count data, the hypothesis under test is that the two sets of counts are independent, but with unspecified marginal distributions. Because the data are discrete, they are naturally presented as a two-way table of frequencies and the hypothesis of independence, conditional on the marginal totals, can be tested via Pearson's X^2 statistic or its asymptotic equivalent, the likelihood ratio statistic for interaction between rows and columns in a Poisson log-linear model. Note that this does not assume that the counts are marginally Poisson-distributed.

For a distance-based approach, an early contribution was made by Goodall (1965) who observed that, under independence, the distances from an arbitrary point to the nearest type 2 event, and from an arbitrary type 1 event to the nearest type 2 event, would be identically distributed. In this context, treating the nearest type 1 event to an arbitrary point as an arbitrary type 1 event would not invalidate a test of independence of the two types of event. It would affect the power of the test, as would switching the labelling of the two types. For example, consider a process in which type 1 events form a Poisson process and a proportion p of type 1 events have an associated type 2 event a small distance away. If p is small, the test as described will be weak, whereas interchanging the role of the two types of event would give a more powerful test since every type 2 event has an associated type 1 event close by, but not conversely.

Diggle and Cox (1983) propose as test statistic Kendall's rank correlation coefficient τ (Kendall, 1970; Sprent, 1981, Chapter 10) between pairs of distances from each of n sample points to the nearest events of type 1 and type 2. Their comparative simulations suggest that this test is more powerful than Goodall's test against a range of positively or negatively dependent alternatives to independence.

3.5 Recommendations

It is difficult to compare the statistical merits of quadrat count and distance methods because they are based on quite different sampling operations, and their relative ease of implementation in the field will vary considerably between applications.

For a quadrat count analysis, the choice of which statistic to use for testing or estimation is straightforward. In contrast, the choice of quadrat size is rather arbitrary and can seriously affect the results of the analysis. With this qualification, the index of dispersion (3.1) provides a test of CSR which is generally powerful against aggregation, but less so against regularity. The intensity estimator $\hat{\lambda}$ defined at (3.3) is always unbiased.

Amongst the many distance-based methods, T-square sampling is easy to use in the field, and the simplicity of its associated distribution theory gives some degree of flexibility in the construction of tests of CSR and estimators of intensity. To test CSR, the combination of t_N and M allows a four-way classification of the underlying pattern as regular, random, heterogeneous or aggregated, with good power against each type of alternative. To estimate the mean area per event, the estimator $\tilde{\gamma}_T$ is generally robust, although unbiasedness is not guaranteed. Catana's method also has the advantages of ease of use in the field and tractable distribution theory under complete spatial randomness. Moreover, it naturally provides a systematic coverage of the region of interest through a series of parallel transects, inviting a more flexible and searching analysis strategy than merely estimating intensity and testing for departure from complete spatial randomnness.

Finally, note that sparse sampling methods are inherently limited in what they can achieve by comparison with methods for the analysis of mapped data. In this respect, we again emphasise that the discussion of the Lansing Woods data in this chapter is purely illustrative.

4

Spatial point processes

CONTENTS

4.1 Processes and summary descriptions

A *spatial point process* is a stochastic mechanism that generates a countable set of events x_i in the plane. We will usually be dealing with processes that are *stationary* and *isotropic*. Stationarity means that all properties of the process are invariant under translation, isotropy that they are invariant under rotation. These assumptions are less restrictive than they might seem at first sight. In particular, they do not rule out the modelling of *random* heterogeneity in the environment (recall the discussion of Figure 1.2 in Section 1.1). However, three qualifications are in order.

Firstly, although models are often defined as processes on the whole plane, in practice we only apply them to data from finite planar regions and it will be sufficient for our purposes if stationarity and isotropy hold to a reasonable approximation within the study region in question. Indeed, study regions are often selected with this requirement implicitly in mind so that, for example, in micro-anatomical studies the tissue sections to be analysed are deliberately sited well away from any boundaries between different types of tissue.

Secondly, we will abandon the stationarity assumption when the data include spatial explanatory variables that account for spatial variation in the local intensity of events.

Finally, in some applications we may choose to circumvent the stationarity assumption by the use of design-based inference. This applies to the analysis of replicated patterns as discussed in Chapter 5, to certain problems in environmental epidemiology which we consider in Chapter 9, and to our discussion of mechanistic models for spatio-temporal point processes in Chapter 13.

Many statistical methods for spatial point pattern data involve comparisons between empirical summary descriptions of the data and the corresponding theoretical summary descriptions of a point process model. However, it is important that the theoretical summary descriptions are derived from an underlying model, rather than being advanced as models in their own right. For example, we have already seen in Chapter 2 that one summary description of the homogeneous Poisson process is that the distribution function of the distance from an arbitrary event of the process to its nearest neighbour is

$$G(y) = 1 - \exp(-\lambda \pi y^2) : y \geq 0.$$

This leads, amongst other things, to the construction of tests of complete spatial randomness involving a comparison between this theoretical form of $G(y)$ and the corresponding empirical distribution function for an observed pattern of n events. However, it would make no sense to attempt to define a more general class of models by embedding $G(y)$ within a larger parametric family of distributions unless the enlarged class were itself derived from an explicit class of point process models.

In this chapter, we consider various theoretical summary descriptions of point processes, and the corresponding empirical descriptions of point pattern data. We focus on properties that lead to useful statistical methods, and illustrate their use on a number of data-sets. We include a description of the homogeneous Poisson process and, for multivariate processes, discuss the ideas of independence and random labelling. We postpone until Chapter 6 a general discussion of different parametric classes of point processes that have been proposed as models for data.

We need the following notation: $E[X]$ denotes the expectation of a random variable X; $N(A)$ denotes the number of events in the planar region A; in a multivariate process, $N_j(A)$ similarly denotes the number of type j events in A; $|A|$ is the area of A; dx is an infinitesimal region that contains the point x; $||x - y||$ denotes the Euclidean distance between the points x and y.

4.2 Second-order properties

4.2.1 Univariate processes

We can now define the *first-order* and *second-order* properties of a spatial point process. First-order properties are described by an *intensity function*,

$$\lambda(x) = \lim_{|dx| \to 0} \left\{ \frac{E[N(dx)]}{|dx|} \right\}.$$

For a stationary process, $\lambda(x)$ assumes a constant value λ, the mean number of events per unit area.

The *second-order intensity function* is similarly defined as

$$\lambda_2(x, y) = \lim_{|dx|, |dy| \to 0} \left\{ \frac{E[N(dx)N(dy)]}{|dx||dy|} \right\}.$$

A closely related quantity is the *conditional intensity* $\lambda_c(x|y) = \lambda_2(x, y)/\lambda(y)$ which, loosely speaking, corresponds to the intensity at the point x conditional on the information that there is an event at y.

For a stationary process, $\lambda_2(x, y) \equiv \lambda_2(x - y)$; for a stationary, isotropic process, $\lambda_2(x - y)$ reduces further to $\lambda_2(t)$, where $t = ||x - y||$. In statistical mechanics, the scaled function $\rho(t) = \lambda_2(t)/\lambda^2$ is referred to as the *radial distribution function*, or *pair correlation function*, although it is neither a distribution function nor a correlation function in the usual statistical sense.

Baddeley, Møller and Waagepetersen (2000) discuss *intensity-reweighted (second-order) stationary* processes, which have the property that

$$\lambda_2(x, y)/\lambda(x)\lambda(y) = \rho(t) \tag{4.1}$$

depends only on $t = ||x - y||$. Note that this requires the intensity function $\lambda(x)$ to be bounded away from zero, in which case we again call $\rho(t)$ the pair correlation function. Intensity-reweighted stationarity is a point process analogue of the assumption commonly made in the analysis of real-valued spatial processes that the mean value may vary spatially whereas the variation about the local mean is stationary.

An alternative characterisation of the second-order properties of a stationary, isotropic process is provided by the function $K(t)$, one definition of which is

$$K(t) = \lambda^{-1}E[N_0(t)], \tag{4.2}$$

where $N_0(t)$ is the number of further events within distance t of an arbitrary event. The notion of an arbitrary event of the process involves the conceptual limit of simple random sampling from a finite population. For a mathematically rigorous discussion see, for example, Daley and Vere-Jones (2002, 2005). Intuitively, we envisage a large but finite number n of events in some finite region and define an arbitrary event to be an event selected at random from this population. Similarly, in practice an arbitrary point will mean a point distributed uniformly over some finite region.

In order to establish a link between $K(t)$ and $\lambda_2(t)$ we shall assume that our process is *orderly*, by which we mean that multiple coincident events cannot occur or, more precisely, that $P\{N(dx) > 1\}$ is of a smaller order of magnitude than $|dx|$. This means that $E[N(dx)] \sim P\{N(dx) = 1\}$ in the sense that the ratio of these two quantities tends to 1 as $|dx| \to 0$. We shall further assume that in a similar sense, $E[N(dx)N(dy)] \sim P\{N(dx) = N(dy) = 1\}$. Under these conditions, the expected number of further events within distance t of an arbitrary event can be computed by integrating the conditional intensity over the disc with centre the origin and radius t. Hence,

$$\lambda K(t) = \int_0^{2\pi} \int_0^t \{\lambda_c(x|o)xdxd\theta.$$

Using the fact that $\lambda_c(x|o) = \lambda_2(x)/\lambda$, this gives

$$\lambda K(t) = 2\pi\lambda^{-1} \int_0^t \lambda_2(x)xdx, \tag{4.3}$$

or conversely, $\lambda_2(t) = \lambda^2(2\pi t)^{-1}K'(t)$ and

$$\rho(t) = (2\pi t)^{-1}K'(t). \tag{4.4}$$

Note that for an intensity-reweighted stationary process, Baddeley, Møller and Waagepetersen (2000) extend the definition of the K-function to

$$K_I(t) = 2\pi \int_0^t \rho(x)xdx,$$

which reduces to (4.3) in the stationary case.

It is sometimes more convenient to work with $\lambda_2(t)$ or $\rho(t)$ rather than with $K(t)$. As a minor variation we define a *covariance density*,

$$\gamma(t) = \lambda_2(t) - \lambda^2 = \lambda^2\{\rho(t) - 1\}. \tag{4.5}$$

For data analysis, one potential advantage of $K(t)$ over $\lambda_2(t)$ or $\rho(t)$, especially in small samples, is that its estimation is more straightforward. Essentially, $K(t)$ and $\lambda_2(t)$ are related to the distribution function and probability density function of the distances between pairs of events in a point pattern, and the former can be estimated without having to decide how much to smooth the corresponding empirical distribution. Against this, given a data-set containing sufficiently many events, a simple histogram-like estimate of $\lambda_2(t)$ or $\rho(t)$ is easily calculated and may be considered easier to interpret. We return to this point in Section 4.6.2.

Another useful property of the K-function is that it is invariant under random thinning. By this, we mean that if each event of a process is retained or not according to a series of mutually independent Bernoulli trials, then the K-function of the resulting thinned process is identical to that of the original, unthinned process. This follows from the definition (4.2), where the K-function is defined as the ratio of two quantities, $E[N_0(t)]$ and λ. The effect of the thinning is to multiply each of these by p, the retention probability for any one event, leaving the ratio unchanged.

Rather than observe the exact locations of events in a planar region, it is sometimes easier in practice to observe only counts $N(A)$ in convenient sub-regions A (cf. Section 2.7). The resulting *quadrat count distribution*,

$$p_n(A) = P\{N(A) = n\} : n = 0, 1, ...,$$

provides a possible summary description of the process. The arbitrary nature of A is unsatisfactory. One solution is further to summarize the quadrat count distribution by its first few moments, and to regard these as functions of A. In particular,

$$E[N(A)] = \int_A \lambda(x)dx,$$

which reduces to $\lambda|A|$ for a stationary process. More interestingly, orderliness implies that

$$\begin{aligned} E[N(A)^2] &= E\left[\left\{\int_A N(dx)\right\}^2\right] \\ &= E\left[\int_A N(dx) + \int_A \int_A N(dx)N(dy)\right]. \end{aligned}$$

Interchanging expectation and integration this becomes, for a stationary process,

$$\lambda|A| + \int_A \int_A \lambda_2(x-y)dxdy,$$

whence

$$\mathrm{Var}\{N(A)\} = \int_A \int_A \lambda_2(x-y)dxdy + \lambda|A|(1 - \lambda|A|). \tag{4.6}$$

A straightforward generalization gives

$$\mathrm{Cov}\{N(A), N(B)\} = \int_A \int_B \lambda_2(x-y)dxdy + \lambda|A \cap B| - \lambda^2|A||B|$$

where $A \cap B$ denotes the intersection of A and B.

The quantity $\mathrm{Var}\{N(A)\}$ defined in (4.6) is sometimes called the "variance-area curve", and is closely related to $K(t)$ in the sense that both involve integrated versions of the second-order intensity function.

4.2.2   Extension to multivariate processes

In a *multivariate* process, the events are of two or more distinguishable types. Definitions for the second-order properties of such processes follow as natural generalisations of the corresponding quantities for univariate processes. We assume stationarity, isotropy and orderliness, and write $N_j(S)$ for the number of type j events in a planar region A.

The *(first-order) intensities* are constants,

$$\lambda_j = E[N_j(A)]/|A|.$$

The *second-order intensities* are functions of a scalar argument,

$$\lambda_{ij}(t) = \lim_{\substack{|dx| \to 0 \\ |dy| \to 0}} \left\{ \frac{E[N_i(dx)N_j(dy)]}{|dx||dy|} \right\}$$

where, as before, t denotes distance. The corresponding *covariance densities* are

$$\gamma_{ij}(t) = \lambda_{ij}(t) - \lambda_i \lambda_j$$

and the multivariate K-functions are

$$K_{ij}(t) = \lambda_j^{-1} E[N_{0ij}(t)] \tag{4.7}$$

where $N_{0ij}(t)$ denotes the expected number of (further) type j events within distance t of an arbitrary type i event. Note that $\lambda_{ij}(t) = \lambda_{ji}(t)$, from which it follows that $\gamma_{ij}(t) = \gamma_{ji}(t)$. A similar argument to the one used in Section 4.2.1 shows that

$$K_{ij}(t) = 2\pi(\lambda_i \lambda_j)^{-1} \int \lambda_{ij}(x)x\,dx$$

from which it follows that $K_{ji}(t) = K_{ij}(t)$.

4.3   Higher order moments and nearest neighbour distributions

Second-order properties provide a natural and valuable starting point for the description of a spatial point process. However, they do not give a complete picture. Baddeley and Silverman (1984) describe a class of non-Poisson processes for which $K(t) = \pi t^2$, and in Section 6.9.3 we shall give another example in which clearly different processes have identical second-order properties. Higher order properties can easily be defined in terms of the joint intensity functions for the occurrence of specified configurations of three, four, etc. events. Interpretation would be difficult in practice since, for example, the third-order intensity function of a stationary, isotropic process requires three arguments, the fourth-order function five, and so on.

In view of this, we define two distribution functions that we used in Chapter 2 to provide tests of CSR and that serve as additional summary descriptions for spatial point processes. These are $G(y)$, the distribution function of the distance from an arbitrary *event* to the nearest other event, and $F(x)$, the distribution function of the distance from an arbitrary *point* to the nearest event.

One interpretation of $F(x)$ is as the probability that a randomly located disc of radius x contains at least one event. This suggests an obvious extension whereby the disc is replaced by some other geometrical figure. Oriented shapes such as ellipses or rectangles could be used to describe departures from isotropy. Matheron (1975) incorporates these ideas within a general theory of random sets, in which a random set S is characterized by the function $\mathcal{F}(\mathcal{T})$, the probability that the intersection of S with T is non-empty, for a suitably wide class of "test-sets " T.

4.4 The homogeneous Poisson process

The homogeneous planar Poisson process, subsequently referred to without qualification as the Poisson process, is the cornerstone on which the theory of spatial point processes is built. It represents the simplest possible stochastic mechanism for the generation of spatial point patterns, and in applications is used as an idealized standard of complete spatial randomness that, if strictly unattainable in practice, sometimes provides a useful approximate description of an observed pattern. The Poisson process is conveniently defined by the following postulates, which correspond exactly to the definition of complete spatial randomness given in Section 1.3:

PP1 For some $\lambda > 0$, and any finite planar region A, $N(A)$ follows a Poisson distribution with mean $\lambda|A|$.

PP2 Given $N(A) = n$, the n events in A form an independent random sample from the uniform distribution on A.

To demonstrate that PP1 and PP2 are self-consistent, we first establish

PP3 For any two disjoint regions A and B, the random variables $N(A)$ and $N(B)$ are independent.

Let $C = A \cup B$ be the union of two disjoint regions A and B. Write $p = |A|/|C|$ and $q = 1 - p = |B|/|C|$. Then, PP2 applied to the region C implies that

$$P\{N(A) = x, N(B) = y | N(C) = n\} = \binom{x+y}{x} p^x q^y,$$

for integers $0 \le x \le n$ and $y = n - x$. PP1 then gives the unconditional joint distribution of $N(A)$ and $N(B)$ as

$$
\begin{aligned}
P\{N(A) = x, N(B) = y\} &= \binom{x+y}{x} p^x q^y \{e^{-\lambda|C|}(\lambda|C|)^n / n!\} \\
&= \{e^{-\lambda|A|}(\lambda|A|)^x / x!\}\{e^{-\lambda|B|}(\lambda|B|)^y / y!\} \quad (4.8)
\end{aligned}
$$

for all integers $x \ge 0$ and $y \ge 0$. This establishes PP3 and shows also that $N(A)$ and $N(B)$ have the distributions implied by PP1. It is immediately obvious that if PP2 holds for any region C it must hold also for all sub-regions of C. Conversely, the additive property of independent Poisson-distributed random variables X and Y, and the associated conditional binomial distribution of X given $X + Y$, establish PP1 and PP2 respectively for any region formed as the union of two disjoint regions for which PP1 and PP2 hold. This proves the required self-consistency.

The parameter λ of the Poisson process is its intensity. The independence result PP3 implies that

$$
\lambda_2(t) = \lambda^2 : t > 0, \tag{4.9}
$$

whence (4.9) gives

$$
K(t) = \pi t^2 : t > 0. \tag{4.10}
$$

The variance-area curve follows directly from PP1 as

$$
\text{Var}\{N(A)\} = \lambda|A|. \tag{4.11}
$$

The nearest neighbour distribution functions $G(y)$ and $F(x)$ are identical, since the existence of an event at a particular point, x_0 say, has no bearing on the distribution of the remaining number of events in a disc with centre x_0. We deduce from PP1 that

$$
F(x) = G(x) = P\{N(\pi x^2) > 0\} = 1 - \exp(-\pi \lambda x^2) : x > 0. \tag{4.12}
$$

To simulate a partial realisation of a Poisson process on A conditional on a fixed value of $N(A)$, we need to generate events independently according to a uniform distribution on A. Awkward shapes of region can be accommodated by simulating the process on a larger region of a more convenient shape, such as a rectangle or disc, and retaining only those events which lie within A. Alternatively, Hsuan (1979) gives an algorithm for the direct generation of events uniformly distributed on an arbitrary polygon.

If $N(A)$ is required to be randomly varying, this same method can of course be preceded by the simulation of $N(A)$ from the appropriate Poisson distribution. In some implementations, the direct simulation of $N(A)$ is a relatively time-consuming step. Lewis and Shedler (1979) propose an alternative method which can be used when A is a rectangle, say $A = (0, a) \times (0, b)$. This is based on the observation that the x-coordinates of events in the infinite strip

$0 \leq y \leq b$ form a *one-dimensional* Poisson process with intensity λb, from which it follows that the differences between successive ordered x-coordinates are independent realizations of an exponentially distributed random variable with distribution function

$$F(v) = 1 - e^{-\lambda b v}, v > 0$$

(see, for example, Cox and Lewis, 1966, Chapter 2). The corresponding y-coordinates are, again independently, uniformly distributed on $(0, b)$. Note that this method automatically generates the events (x_i, y_i) in order of increasing x-coordinates, and terminates when the latest x-coordinate is greater than a.

4.5   Independence and random labelling

To assess the spatial association between the two types of events in a bivariate process, we can consider at least two different benchmark hypotheses :

(i) *independence* - the two types of event are generated by a pair of independent univariate processes;

(ii) *random labelling* - the two types of event are generated by labelling the events of a univariate process in a series of mutually independent Bernoulli trials.

These two hypotheses generate distinctively different K_{12}-functions.

Firstly, for any two independent processes of type 1 and type 2 events

$$K_{12}(t) = \pi t^2.$$

This follows from the fact that, if the two component processes are independent, then an event of type 1 has the same status, with respect to events of type 2, as an arbitrary point, hence the expected number of type 2 events within a disc of radius t centred on an arbitrary type 1 event is $\lambda_2 \pi t^2$, the expected number of type 2 events per unit area multiplied by the area of the disc. It follows that $K_{12}(t) = \pi t^2$ as claimed.

Secondly, for any randomly labelled process of type 1 and type 2 events,

$$K_{11}(t) = K_{22}(t) = K_{12}(t) = K(t), \tag{4.13}$$

where $K(s)$ is the K-function for the unlabelled univariate process. To see this, let $K(t)$ be the K-function of the unlabelled, univariate process consisting of all events, irrespective of type. Then, under random labelling the univariate processes of type 1 and type 2 events are each random thinnings of the unlabelled process, and we have already seen in Section 4.2.1 that the K-function is invariant under random thinning, hence $K_{11}(t) = K_{22}(t) = K(t)$. Essentially the same argument shows that $K_{12}(t) = K(t)$.

Note that independence and random labelling are equivalent if and only if the component processes of type 1 and type 2 events are both Poisson processes. This makes it important to decide which, if either, is the natural benchmark of "no association" in a particular application.

4.6   Estimation of second-order properties

4.6.1   Stationary processes

For the reason given in Section 4.2, we shall focus initially on estimating the K-function.

In Section 4.2 we defined the function $K(t)$ by

$$\lambda K(t) = E[N_0(t)],$$

the expected number of further events within distance t of an arbitrary event, where the intensity λ is the mean number of events per unit area. An obvious estimator for λ is the observed number of events per unit area, $\hat{\lambda} = n/|A|$.

Similarly, because $E(t) = \mathrm{E}[N_0(t)]$ is the expected number of further events within distance t of an arbitrary event, we can construct an estimator for $E(t)$ as follows. Let $r_{ij} = ||x_i - x_j||$. Define

$$\tilde{E}(t) = n^{-1} \sum_{i=1}^{n} \sum_{j \neq i} I(r_{ij} \leq t), \tag{4.14}$$

where $I(\cdot)$ denotes the indicator function.

The form of the estimator $\tilde{E}(t)$ in (4.14) suggests, correctly, that the K-function is closely connected to the distribution of inter-event distances, whose use in exploratory analysis we discussed in Section 2.2. However, $\tilde{E}(t)$ is negatively biased for $E(t)$ because of edge-effects. For a reference event within distance t of the boundary of A, the observed count of other events within distance t necessarily excludes any events which may have occurred within distance t but outside A. Several methods have been proposed to correct for this source of bias; see, for example, Stein(1991) or Baddeley (1999). The following method, which we shall use in all of our examples, is due to Ripley (1976).

Let $w(x, r)$ be the proportion of the circumference of the circle with centre x and radius r which lies within A. Write w_{ij} for $w(x_i, ||x_i - x_j||)$. Then, for any stationary isotropic process, w_{ij} is the conditional probability that an event is observed, given only that it is a distance $r_{ij} = ||x_i - x_j||$ away from

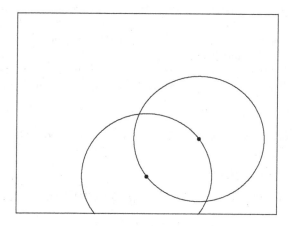

FIGURE 4.1
Construction of the edge-correction weights in Ripley's (1976) estimator for $K(t)$

the ith event x_i. See Figure 4.1, and note that in general $w_{ij} \neq w_{ji}$. Thus, an unbiased estimator for $E(t)$ is

$$\hat{E}(t) = n^{-1} \sum_{i=1}^{n} \sum_{j \neq i} w_{ij}^{-1} I(r_{ij} \leq t).$$

Finally, replacing the unknown intensity λ by $(n-1)/|A|$, we obtain Ripley's (1976) estimator for $K(t)$,

$$\hat{K}(t) = \{n(n-1)\}^{-1} |A| \sum_{i=1}^{n} \sum_{j \neq i} w_{ij}^{-1} I(r_{ij} \leq t). \qquad (4.15)$$

In fact, Ripley used n^{-2} rather than $\{n(n-1)\}^{-1}$ in the expression for $\hat{K}(t)$; we prefer the given form for technical reasons although the distinction is clearly unimportant when n is large.

Ripley's estimator is approximately unbiased for sufficiently small t, the restriction on t being necessary because the weights w_{ij} can become unbounded as t increases. In practice this is not a serious problem. For example, when A is the unit square the theoretical upper limit of t is $\frac{1}{2}\sqrt{2} \approx 0.7$ but $\hat{K}(t)$ will seldom be required for such large values of t, partly because the sampling fluctuations in $\hat{K}(t)$ increase with t but also because it is not realistic to attempt to model effects which operate on the same physical scale as the dimensions of A.

The **splancs** package incorporates an algorithm written by Barry Rowlingson for computing $w(x, y)$ when A is an arbitrary polygon. Explicit formulae for $w(x, u)$ can also be written down for simple shapes of region A, for example rectangular or circular, and these may be useful if computational efficiency is paramount. Suppose firstly that A is the rectangle $(0, a) \times (0, b)$. Write $x = (x_1, x_2)$ and let $d_1 = \min(x_1, a - x_1)$ $d_2 = \min(x_2, b - x_2)$; thus d_1 and d_2 are the distances from the point x to the nearest vertical and horizontal edges of A. To calculate $w(x, u)$ we need to distinguish two cases:

1. if $u^2 \leq d_1^2 + d_2^2$, then

$$w(x, u) = 1 - \pi^{-1}[\cos^{-1}\{\min(d_1, u)/u\} + \cos^{-1}\{\min(d_2, u)/u\}; \tag{4.16}$$

2. if $u^2 > d_1^2 + d_2^2$, then

$$w(x, u) = 0.75 - (2\pi)^{-1}\{\cos^{-1}(d_1/u) + \cos^{-1}(d_2/u)\}. \tag{4.17}$$

Note that (4.16) correctly gives $w(x, u) = 1$ when $u \leq \min(d_1, d_2)$. The above formulae apply to values of u in the range $0 \leq u \leq 0.5\min(a, b)$ which, as noted above, should be sufficient for practical purposes.

Now suppose that A is the disc with centre the origin and radius a. Let $r = \sqrt{(x_1^2 + x_2^2)}$ be the distance from x to the centre of the disc. Then, again distinguishing two cases, we have the following:

1. if $u \leq a - r$, then

$$w(x, u) = 1;$$

2. if $u > a - r$, then

$$w(x, u) = 1 - \pi^{-1}\cos^{-1}\{(a^2 - r^2 - u^2)/(2ru)\}.$$

These formulae apply to values of u between 0 and a.

The sampling distribution of $\hat{K}(t)$ is analytically intractable, except in the case of a homogeneous Poisson process. Given any specific model and region A, the sampling distribution can be estimated by direct simulation. However, the theoretical expression for the variance of $\hat{K}(t)$ in a homogeneous Poisson process provides a useful benchmark in initial inspection of a plot of $\hat{K}(t)$. In what follows, we treat n, the number of events in A, as fixed.

For a homogeneous Poisson process, Ripley (1988) gives an asymptotic approximation to the sampling variance of $\hat{K}(t)$. Lotwick and Silverman (1982) give exact formulae whose evaluation in general requires extensive numerical integration, although they give explicit formulae for rectangular A. Chetwynd and Diggle (1998) give a different approximation, based on a thinning argument, that is easily computed for arbitrarily shaped A.

Ripley's asymptotic approximation, modified to take account of our non-standard choice of denominator in (4.15), is

$$v_R(t) = 2\{|A|/(n-1)\}^2\{\pi t^2/|A| + 0.96Pt^3/|A|^2 + 0.13(n/|A|)Pt^5/|A|^2\} \tag{4.18}$$

where P denotes the perimeter of A. The approximation is accurate for relatively small values of t (see below).

Lotwick and Silverman's result, similarly modified and assuming rectangular A, is that the variance of $\hat{K}(t)$ is

$$v_{LS}(t) = \{n(n-1)\}^{-1}\{2b(t) - a_1(t) + (n-2)a_2(t)\} \qquad (4.19)$$

where, for rectangular A with perimeter length P,

$$b(t) = \pi t^2 |A|^{-1}(1 - \pi t^2/|A|) + |A|^{-2}(1.0716Pt^2 + 2.2375t^4),$$

$$a_1(t) = |A|^{-2}(0.21Pt^3 + 1.3t^4)$$

and

$$a_2(t) = |A|^{-3}(0.24Pt^5 + 2.62t^6).$$

These expressions are valid for t less than or equal to the shorter side-length of A.

Chetwynd and Diggle's approximation involves summations of functions of the edge-correction weights w_{ij} as follows. For any fixed t, define $\phi_{ij} = 0.5(w_{ij} + w_{ji})I(||x_i - x_j|| \leq t)$. Further define

$$W_n = \sum_{i=1}^{n}\sum_{j\neq i} \phi_{ij},$$

$$X_n = \sum_{i=1}^{n}\sum_{j\neq i} \phi_{ij}^2$$

and

$$Z_n = \sum_{i=1}^{n}(\sum_{j\neq i} \phi_{ij})^2.$$

Write $n^{(k)} = n(n-1)...(n-k+1)$ and define $m_2(t) = X_n/n^{(2)}$, $m_3(t) = (Z_n - X_n)/n^{(3)}$ and $m_4(t) = (W_n^2 - 4Z_n + 2X_n)/n^{(4)}$. Then the estimated variance of $\hat{K}(t)$ is

$$v_{CD}(t) = (2|A|^2/n^{(2)})\{(3-2n)m_4(t) + 2(n-2)m_3(t) + m_2(t)\}. \qquad (4.20)$$

Chetwynd and Diggle in fact give an explicit formula for $\text{Cov}\{\hat{K}(t), \hat{K}(s)\}$, of which (4.20) is a special case.

It is also useful to be able to assess the precision of an estimate of $K(t)$ without assuming a specific model. A simple way to do this is to subdivide A into equal sub-areas, estimate $K(t)$ separately within each sub-area and use the empirical variance over the separate estimates. Thus, if for each t we let k_i denote the estimate of $K(t)$ from the ith of m sub-areas, then our overall estimate is

$$\tilde{K}(t) = m^{-1}\sum_{i=1}^{m} k_i \qquad (4.21)$$

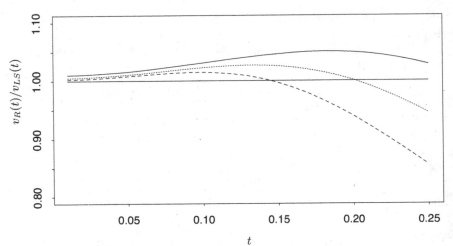

FIGURE 4.2
Comparison between Ripley's asymptotic approximation and the Lotwick-Silverman formula for the sampling variance of $\hat{K}(t)$. ——— : $n = 100$; — — — : $n = 200$; .... : $n = 400$. The horizontal line at height 1 corresponds to equality of the two formulae.

with approximate variance

$$\text{Var}\{\tilde{K}(t)\} \approx \{(m(m-1)\}^{-1}\sum_{i=1}^{m}\{k_i - \tilde{K}(t)\}^2. \qquad (4.22)$$

The approximation in (4.22) arises for two reasons. Firstly, an element of approximation is inherent in using the sample variance of the k_i as an estimate of their true variance; secondly, dividing the sample variance of the k_i by m makes the implicit assumption that disjoint sub-regions give independent estimates k_i, which is correct for the homogeneous Poisson process, but not more generally. Furthermore, $\tilde{K}(t)$ can be expected to be less efficient than $\hat{K}(t)$ because it does not use information from pairs of events in different sub-regions; this may be an important consideration unless n, the number of events, is very large. All of these considerations suggest that the estimator $\tilde{K}(t)$ and its associated approximate variance should be used only for relatively small values of t, or when the artificial subdivision of A is replaced by genuine replication. We postpone further discussion of replicated patterns until Chapter 8.

We now illustrate the performance of the different estimators for $\text{Var}\{\hat{K}(s)\}$. We consider first the homogeneous Poisson process. Figure 4.2 shows the ratio of Ripley's asymptotic approximation (4.18) and the Lotwick-Silverman formula (4.19), for $n = 100$, 200 and 400, and A the unit square. For small t, the approximation is excellent when $n = 100$ and, as would be expected, improves as n increases. At larger values of t, it is less reliable

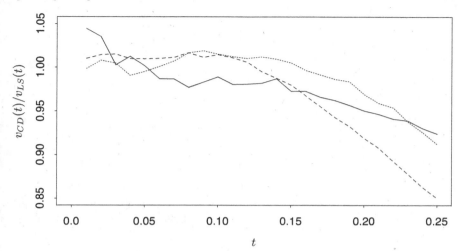

FIGURE 4.3
Comparison between Chetwynd and Diggle's estimator and the Lotwick-Silverman formula for the sampling variance of $\hat{K}(t)$. ——— : $n = 100$; — — — : $n = 200$; .... : $n = 400$.

and does not necessarily improve as n increases. For a comparison between the Lotwick-Silverman formula and the Chetwynd-Diggle method, we need to simulate replicate patterns because the latter is a data-based estimate rather than a numerical approximation. Figure 4.3 is comparable to Figure 4.2 except that the comparison is now between the Chetwynd-Diggle estimated variance, averaged over replicate simulations, and the Lotwick-Silverman formula. The results point to a small negative bias in the Chetwynd-Diggle estimator at large distances.

In a second experiment, we again simulated homogeneous Poisson processes but now compared the two estimators $\hat{K}(t)$ and $\tilde{K}(t)$, the latter using a sub-division of the unit square into a 4×4 grid of smaller squares. The results summarised below are based on 1000 replicates of processes with each of $n = 100$, 200 and 400 events on the unit square, and for distances $t \leq 0.25$.

Recall that for the homogeneous Poisson process, the implicit assumption in (4.22) that disjoint sub-regions give independent estimates of $K(t)$ is correct, and the simulation results confirm this. Perhaps more interestingly, Figure 4.4 shows the ratio of the estimated variance of $\tilde{K}(t)$ and the Lotwick-Silverman formula $v_{LS}(t)$ for the variance of $\hat{K}(t)$. This confirms the inherent inefficiency of $\tilde{K}(t)$ relative to $\hat{K}(t)$, especially for small n and/or large t.

Finally, we simulated two non-Poisson processes, one generating aggregated patterns similar to Figure 1.2, the other regular patterns with a minimum permissible distance between events, similar to Figure 1.3. The results confirm that, as expected, formulae based on a Poisson assumption tend to

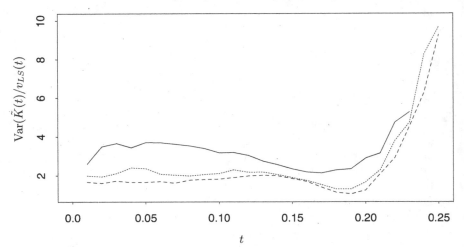

FIGURE 4.4
Comparison between the variances of $\tilde{K}(t)$ and of $\hat{K}(t)$. ——— : $n = 100$; ....
: $n = 200$; – – – : $n = 400$

under-estimate or over-estimate the variance of $\hat{K}(t)$ according to whether the underlying process is aggregated or regular, respectively. Specifically, in the aggregated case the estimated variances were between two and 12 times larger than the Poisson-based Lotwick-Silverman formulae, and in the regular case between three and seven times smaller over the range $t \leq 0.25$ (excluding small distances for which $\hat{K}(t)$ is identically zero because of the minimum permissible inter-event distance). However, the approximation (4.22) continued to give reasonable estimates for the variance of the less efficient estimator $\tilde{K}(t)$.

Our overall conclusions from these comparisons are the following. For the initial exploration of second-order structure of patterns which are close to completely random, it is useful to supplement a graphical display of $\hat{K}(t)$ with error bounds based on the Lotwick-Silverman formula for rectangular regions. For regions with an irregular boundary, such as arise typically in epidemiological applications, either the Chetwynd-Diggle formula can be used, or empirical bounds can be constructed from repeated simulations of a homogeneous Poisson process.

For patterns markedly different from completely random, Poisson-based approximations are unreliable. However, provided that a parametric model has been formulated, standard errors can again be estimated empirically from repeated simulations of the declared model. It is usually instructive, and straightforward in most modern computing environments, to supplement a graphical display of $\hat{K}(t)$ with envelopes of repeated simulations of a candidate model, both to quantify uncertainty and for an informal assessment of goodness-of-fit. We give examples in later chapters.

In the absence of a parametric model, the estimator $\tilde{K}(t)$ can be used if a reliable indication of precision takes precedence over efficiency of estimation.

4.6.2 Estimating the pair correlation function

Given an estimate $\hat{K}(t)$, a simple way to estimate the pair correlation function is to exploit the relationship between $K(t)$ and $\rho(t)$ by substituting the derivative of $\hat{K}(t)$ for $K'(t)$ on the right hand side of (4.4). Strictly, $\hat{K}(t)$ as defined at (4.15) is a step-function, and therefore non-differentiable. However, in practice we usually evaluate $\hat{K}(t)$ at a discrete set of equally spaced values of t, say $t = jh : j = 0, 1, 2, ..., m$ for a suitably small value $h > 0$ and interpolate linearly. In this case, $\hat{K}'(t)$ is piece-wise constant,

$$\hat{K}'(t) = \{\hat{K}(jh) - \hat{K}((j-1)h)\}/h :: t \in ((j-1)h, jh),$$

leading to a histogram-like estimator for $\rho(t)$,

$$\hat{\rho}(t) = (2\pi t)^{-1}\hat{K}'(t). \tag{4.23}$$

Stoyan and Stoyan (1994) discuss a kernel-smoothed version, subsequently used by a number of authors including Møller, Syversveen and Waagepetersen (1998). For this, we again choose a positive value h, but also specify a univariate probability density function, $g(u)$ say, symmetric about $u = 0$, to define the smoothing weights. Stoyan and Stoyan's estimator for $\rho(t)$ is then

$$\tilde{\rho}(t) = |A|(2\pi t n^2)^{-1} \sum_{j \neq i} w_{ij}^{-1} h^{-1} g\{(t - u_{ij})/h\} \tag{4.24}$$

where, as earlier, $|A|$ is the area of A, $u_{ij} = ||x_i - x_j||$ and w_{ij} is Ripley's edge-correction. In the non-parametric smoothing literature, h is called the *band-width* and $g(u)$ the *kernel function*; see, for example, Silverman (1986).

Typically, h in (4.24) would usually be chosen somewhat larger than its counterpart when estimating $K(t)$. Another consideration is that (4.24) gives poor estimates of $\rho(t)$ at values of t close to zero; indeed, it gives positive estimates for negative t. A simple solution to this is to reflect the negative portion of the estimate in the y-axis, i.e. for each positive t, use the estimate

$$\hat{\rho}(t) = \tilde{\rho}(t) + \tilde{\rho}(-t), \tag{4.25}$$

although this can still leave substantial bias near $t = 0$ for aggregated patterns. Note also that the reflection in the y-axis does not arise for the histogram-like estimator (4.23), although the tendency to under-estimate $\rho(t)$ near $t = 0$ for aggregated patterns remains.

To illustrate these issues, we generated synthetic data by simulating a Poisson process with intensity $n = 500$ events on a unit square region A.

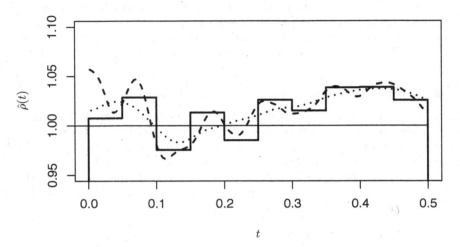

FIGURE 4.5

Comparison between histogram-like (solid line) and kernel smoothed estimates of the pair correlation function. Dashed and dotted lines correspond to kernel estimates using a quartic kernel and band-widths $h = 0.038$ and 0.076, respectively.

Figure 4.5 shows estimates (4.23) and two versions of (4.25), based on the simulated data, and using a quartic kernel function,

$$g(u) = (15/16)(1 - u^2)^2 : -1 < u < 1,$$

in the smoothed estimator defined by (4.24) and (4.25). The histogram-like estimate uses bins of width 0.05, whilst the two versions of the kernel estimate use band-widths $h = 0.038$ and 0.076. The smaller of the two band-widths induces an amount of smoothing comparable to the histogram-like estimate except near $u = 0$, apart from which the difference between these two is largely cosmetic, whereas the larger band-width gives a materially smoother estimate. Stoyan and Stoyan (1994) offer guidelines for choosing the band-width h in (4.24). The author's opinion is that attempting to optimise the choice of band-width is an over-elaboration of what is best seen as a simple exploratory device.

4.6.3   Intensity-reweighted stationary processes

If we assume that the process is intensity-reweighted stationary and, unrealistically, that the first-order intensity $\lambda(x)$ is known, we can estimate $K_I(t)$ by an easy modification of (4.18). The modification consists of re-scaling the inter-event distances by the product of the first-order intensities at the corresponding two locations, leading to the estimator

$$\hat{K}_I(t; \lambda) = |A|^{-1} \sum_{i=1}^{n} \sum_{j \neq i} w_{ij}^{-1} I_t(u_{ij}) / \{\lambda(x_i)\lambda(x_j)\}. \qquad (4.26)$$

This was proposed by Baddeley, Møller and Waagepetersen (2000), who went on to discuss the consequences of using an estimated first-order intensity $\hat{\lambda}(x)$ in place of the true $\lambda(x)$. Unsurprisingly, this turns out to be problematic, because of the difficulty of distinguishing empirically between non-constancy of $\lambda(x)$ and dependence between the events of the process. This relates to the equivalence of certain classes of Cox process and Poisson cluster process, to be discussed in Chapter 6. As a consequence, it is difficult in practice simultaneously to estimate non-parametrically both first-order and second-order properties of an intensity-reweighted stationary process. One situation in which it is easier to disentangle first-order and second-order properties is in design-based inference when independent replicate patterns can be assumed to have the same first-order intensity function. Another is in the analysis of case-control data in epidemiology. We discuss these topics in Chapters 5 and 9, respectively.

4.6.4 Multivariate processes

To estimate $K_{12}(s)$ for a bivariate pattern we use the same basic idea as in estimating $K(s)$, but measure distances between pairs of events of different types. Thus, if u_{ij} is the distance between the ith event of type 1 and the jth event of type 2, w_{ij} is as before, and the numbers of type 1 and type 2 events are n_1 and n_2 respectively, we can construct two estimates of $\hat{K}_{12}(s)$ as follows:

(i) $\hat{\lambda}_2 \tilde{K}_{12}(s) = n_1^{-1} \sum_{i=1}^{n_1} \sum_{j=1}^{n_2} w_{ij} I(u_{ij} \leq s)$

(ii) $\hat{\lambda}_1 \tilde{K}_{21}(s) = n_2^{-1} \sum_{j=1}^{n_2} \sum_{i=1}^{n_1} w_{ji} I(u_{ij} \leq s)$

We then combine the two estimates as a weighted average, to give

$$
\begin{aligned}
\hat{K}_{12}(s) &= (n_1 n_2)^{-1} |A| \left\{ n_1 \sum_{i=1}^{n_1} \sum_{j=1}^{n_2} w_{ij} I(u_{ij} \leq s) \right. \\
&\quad + \left. n_2 \sum_{j=1}^{n_2} \sum_{i=1}^{n_1} w_{ji} I(u_{ij} \leq s) \right\} / (n_1 + n_2) \\
&= (n_1 n_2)^{-1} |A| \sum_{i=1}^{n_1} \sum_{j=1}^{n_2} w_{ij}^* I(u_{ij} \leq s), \qquad (4.27)
\end{aligned}
$$

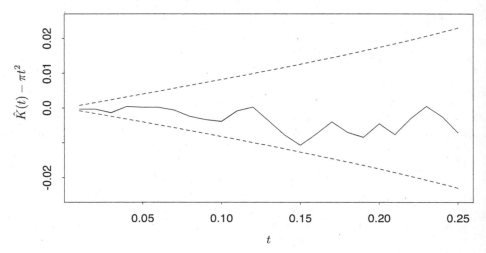

FIGURE 4.6
The estimate $\hat{K}(t) - \pi t^2$ for the Japanese black pine data. ——— : data; — — — : plus and minus two standard errors under complete spatial randomness.

where
$$w_{ij}^* = (n_1 w_{ij} + n_2 w_{ji})/(n_1 + n_2).$$

The variance formulae in Lotwick and Silverman (1982) include the bivariate case. In addition to the functions $v(t)$, $a_1(t)$ and $a_2(t)$ defined above, let $c = n_2/(n_1 + n_2)$. Then, when the component processes are independent homogeneous Poisson processes,

$$\text{Var}\{\hat{K}_{12}(t)\} = (n_1 n_2)^{-1}|A|^2[v(t) - 2c(1-c)a_1(t) + \{(n_1 - 1)c^2 + (n_2 - 1)(1-c)^2\}a_2(t)] \quad (4.28)$$

4.6.5 Examples

Figure 4.6 shows the estimate $\hat{D}(t) = \hat{K}(t) - \pi t^2$ for the Japanese black pine sapling data of Figure 1.1, together with plus and minus two standard deviation limits calculated from the Lotwick-Silverman formula under the assumption that the data are generated by a homogeneous Poisson process. Note that $\hat{D}(t)$ lies within these limits throughout the plotted range, suggesting compatibility with the Poisson assumption as in earlier analyses of these data.

Note also that the standard deviation under the Poisson assumption is roughly linear in t. This is an indirect consequence of the Poisson quadrat count distribution, since $\hat{K}(t)$ is essentially an average of counts in circles of radius t, hence its mean and variance are both approximately proportional

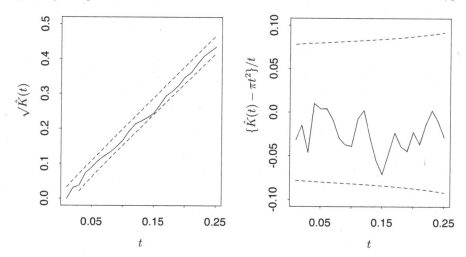

FIGURE 4.7
Transformed estimates of $\hat{K}(t)$ for the Japanese black pine data. —— : data;
— — — : plus and minus two standard errors under complete spatial randomness.
The left-hand panel shows $\sqrt{\hat{K}(t)}$, the right-hand panel $\{\hat{K}(t) - \pi t^2\}/t$.

to t^2. For this reason, some authors recommend plotting $\sqrt{\hat{K}(t)}$ against t to
stabilise the sampling variance, and incidentally to linearise the plot under
the Poisson assumption. We prefer to plot $\hat{D}(t)$ because of its direct physical
interpretation in terms of counting numbers of events in circular regions; also,
as we shall show in later chapters, plots of $D(t)$ can be used for preliminary
estimation of parameters for some widely used models. If a variance-stable
plot is required, either the square-root scale can be used, or a standardised
difference, $\hat{D}(t)/t$, although the latter would become numerically unstable if
extrapolated to $t = 0$. The two panels of Figure 4.7 show these two plots for
the Japanese black pine data. In each case, the error limits are obtained by the
appropriate transformation of the plus and minus two standard error limits
for $\hat{K}(t)$ calculated by the Lotwick-Silverman formula.

The two panels of Figure 4.8 show $\hat{D}(t)$ with plus and minus two standard
error limits under complete spatial randomness, for the redwood data of Figure
1.2, and for the cell data of Figure 1.3. In contrast to Figure 4.6, it is clear in
both cases that the data are incompatible with complete spatial randomness,
but for opposite reasons. As is now very familiar, the redwood data display
strong spatial aggregation, and the cell data strong spatial regularity. The
damped oscillatory behaviour of $\hat{D}(t)$ for the cell data is typical of regular
patterns.

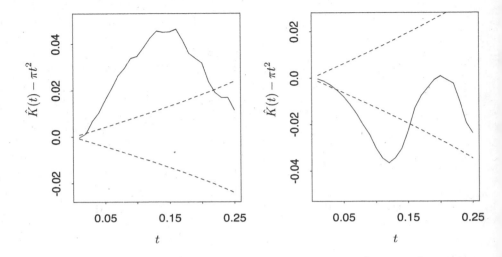

FIGURE 4.8
The estimate $\hat{D}(t) = \hat{K}(t) - \pi t^2$ for the redwood data (left-hand panel) and for the cell data (right-hand panel). ——— : data; — — — : plus and minus two standard errors under complete spatial randomness.

4.7   Displaced amacrine cells in the retina of a rabbit

The displaced amacrine cell data, shown in Figure 1.4, illustrate how estimated K-functions can be used in a specific context without explicit parametric modelling. The analysis presented here is adapted from Diggle (1985a).

The primary scientific interest in these data is to distinguish between two developmental hypotheses (Hughes, 1981). Recall that the two types of cell are those which respond to a light being switched *on* or *off*, respectively. The *separate layer* hypothesis is that the on and off cells are initially formed in two separate layers which later fuse to form the mature retina, whilst the *single layer* hypothesis is that the two types of cell are initially undifferentiated in a single layer and acquire their separate functions at a later stage.

Figure 4.9 shows estimates of $K_{ij}(t)$ for each of $(i, j) = (1, 1)$, $(1, 2)$ and $(2, 2)$, where types 1 and 2 refer to on and off cells, respectively, and an estimate of $K(t)$ for the superposition of both types of cell. Note firstly that the estimate of $K_{12}(t) - \pi t^2$ is close to zero throughout the plotted range. We could use the Lotwick-Silverman formula to compute the sampling variance of $K_{12}(t)$ under the assumption that the two types of cell form independent Poisson processes, but this would likely over-estimate the variance because, as shown by the estimates of $K_{11}(t)$ and $K_{22}(t)$, the components' patterns are markedly more regular than the Poisson process. However, we can test the

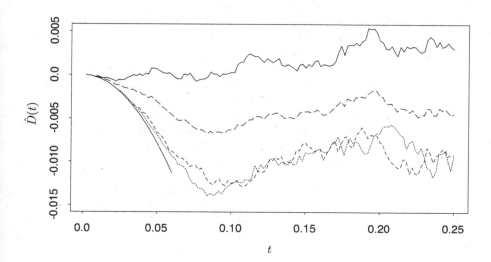

FIGURE 4.9

Second-order properties of displaced amacrine cells. Functions plotted are $\hat{D}(t) = \hat{K}(t) - \pi t^2$ as follows: $- - -$: on cells; .... : off cells; $— — —$: all cells; ——— : bivariate. The parabola $-\pi t^2$ is also shown as a solid line.

independence hypothesis using a procedure also suggested by Lotwick and Silverman (1982). If we wrap the rectangular observation window A onto a torus, independence of the component processes would imply that the sampling distribution of $\hat{K}_{12}(t)$ is invariant to a random toroidal shift of either pattern. This in turn implies that we can conduct a Monte Carlo test of independence by comparing the value of a suitable test statistic for the observed data with values generated under a sequence of independent random toroidal shifts. Using the test statistic

$$u = \sum_{k=1}^{125} t_k^{-2} \{\hat{K}_{12}(t_k) - \pi t_k^2\}^2,$$

where $t_k = 0.002, 0.004, ..., 0.250$, we implemented a Monte Carlo test by re-calculating u after each of 99 independent random toroidal shifts, and obtained a p-value of 0.12. Thus, at least with regard to their second-order properties, the evidence against the hypothesis of independent components is weak. This is supportive of the separate layer hypothesis. However, the evidence is not yet conclusive since the result that $K_{12}(t) = \pi t^2$ is necessary, but not sufficient, for independence.

We now examine $\hat{K}_{11}(t)$ and $\hat{K}_{22}(t)$, and note that they are very similar to each other, but markedly different from $\hat{K}(t)$ for the superposition. This suggests that if the observed pattern is the result of a labelling of an initially undifferentiated set of cells, then the labelling cannot be random. We could formally test the hypothesis of random labelling, but this seems unnecessary in view of the very clear difference between $\hat{K}(t)$ for the superposition and the two very similar estimates $\hat{K}_{11}(t)$ and $\hat{K}_{22}(t)$.

The argument so far leaves open the possibility that the observed pattern is the result of a non-random labelling process. However, note that both $\hat{K}_{11}(t)$ and $\hat{K}_{22}(t)$ are zero for $t < 0.025$ approximately, whereas $\hat{K}(t)$ for the superposition is non-zero at much smaller values of t. For this to arise from a labelling process, the following would have to be true: in the undifferentiated process, close pairs of cells are allowed but mutually close triples are forbidden; and in the labelling process, the two members of a close pair must always be oppositely labelled. This seems implausible, and the analysis therefore points strongly towards the separate layer hypothesis being the correct explanation. Subsequent work by Eglin, Diggle and Troy (2005) and Diggle, Eglen and Troy (2006) led to a refinement of the separate layer hypothesis; in Section 8.3.6 we shall re-visit this example accordingly.

4.8   Estimation of nearest neighbour distributions

We now consider estimation of the two nearest neighbour distribution functions introduced in Sections 2.3 and 2.4. These are: $F(x)$, the probability that the distance from an arbitrary point to the nearest event is less than or equal to x; and $G(x)$, the probability that the distance from an arbitrary event to the nearest other event is less than or equal to x.

In either case, the simplest estimator is the empirical distribution function of observed nearest neighbour distances, as used in Sections 2.3 and 2.4. These estimators are biased because of edge-effects. The bias does not affect the validity of a Monte Carlo test of complete spatial randomness, or indeed of any other specified model, but would be problematic if we were directly concerned with estimation.

One approach to edge-correction is the following, due to Ripley (1977). Consider first an estimator for $G(y)$. Let $(y_i, d_i) : i = 1, ..., n$, denote the distances from each event to the nearest other event in A and to the nearest point on the boundary of A, respectively, and define the estimator $\tilde{G}(y)$ to be the proportion of nearest neighbour distances $y_i \leq y$ amongst those events at least a distance y from the boundary of A; thus

$$\tilde{G}(y) = \#(y_i \leq y, d_i > y)/\#(d_i > y).$$

To estimate $F(x)$, similarly let $(x_i, e_i), i = 1, \ldots, m$, denote the distances from each of m sample points to the nearest event in A and to the nearest point on the boundary of A, and define

$$\tilde{F}(x) = \#(x_i \leq x, e_i > x)/\#(e_i > x).$$

As in Section 2.4, for estimation of $F(x)$ we would suggest locating the m sample points in a regular grid. Alternatively, we could avoid altogether the need for a grid of sample points by computing directly the areal proportion of the study region for which the distance to the nearest event is less than or equal to x, as described by Lotwick (1981).

A technically simpler procedure for dealing with edge-effects would be to use the buffer-zone method, taking measurements only from points or events sufficiently far from the boundary of A, so that edge effects do not arise for the range of distances of interest. As noted in Section 1.3, the obvious disadvantage of this procedure is that it effectively throws away a non-trivial proportion of the data. Recall also from Section (2.7) that Baddeley and van Lieshout's J-function, defined by $J(x) = \{1 - G(x)\}/\{1 - F(x)\}$, can be estimated reliably without any correction for edge-effects.

The sampling distributions of estimators for $F(\cdot)$ and $G(\cdot)$ appear to be intractable, although for large data-sets the device of splitting the observation region A into sub-regions to form pseudo-replicates is available, whilst simulation can be used to assess the fit of a parametric model.

4.8.1 Examples

Figure 4.10 compares the empirical distribution functions, $\hat{G}(\cdot)$ and $\hat{F}(\cdot)$, with the edge-corrected estimators $\tilde{G}(\cdot)$ and $\tilde{F}(\cdot)$ for the Japanese black pine sapling data of Figure 1.1. Notice that the two estimates in each case are very similar, but that the edge-corrected estimators are not necessarily monotone.

Figures 4.11 and 4.12 are the corresponding plots for the redwood seedling data of Figure 1.2 and for the cell data of Figure 1.3. For comparability, all six plots in Figures 4.10, 4.11 and 4.12 evaluate the estimates over the same set of distances. The qualitative differences amongst these three data-sets are again clear.

4.9   Concluding remarks

The K-function, and the two nearest neighbour distribution functions $F(\cdot)$ and $G(\cdot)$, provide complementary tools for the description of spatial point processes.

The K-function is the most amenable of the three to theoretical analysis. We shall see in later chapters that its algebraic form can be derived for a number of useful models. Its physical interpretation as a scaled expectation is also a useful property. The sampling distribution of its estimator $\hat{K}(t)$ is reasonably well understood in the case of a homogeneous Poisson process,

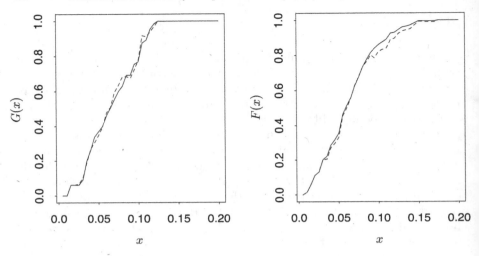

FIGURE 4.10
Estimates of $F(x)$ (left-hand panel) and of $G(x)$ (right-hand panel) for the Japanese black pine data. —— : $\hat{G}(\cdot), \hat{F}(\cdot)$; – – – : $\tilde{G}(\cdot), \tilde{F}(\cdot)$

but otherwise must be assessed by splitting the observation region into sub-regions, or by simulation. The choice between using estimates of $K(t)$ or $\rho(t)$ to summarise the second-order properties of an observed pattern is to some extent a matter of taste. One clear advantage in estimating $K(t)$ rather than $\rho(t)$ is that this avoids the need to choose a bin-width or band-width. In the author's opinion, this is conclusive when the number of events is relatively small, say of the order of one hundred or less. For observed patterns with larger numbers of events, the choice is less clear-cut.

Estimates of all three functions, $K(\cdot)$ (or $\rho(\cdot)$), $F(\cdot)$ and $G(\cdot)$ are useful for exploratory assessment of departure from complete spatial randomness. More generally, we shall see in later chapters that the estimated K-function is a useful and versatile tool. In the author's opinion, the principal use of estimates of $F(\cdot)$ and $G(\cdot)$ is to provide goodness-of-fit measures that are complementary to the second-order description provided by the K-function. In this respect, it is worth noting that the function $J(x) = \{1-G(x)\}/\{1-F(x)\}$ can sometimes be calculated explicitly when its separate components $F(x)$ and $G(x)$ are intractable.

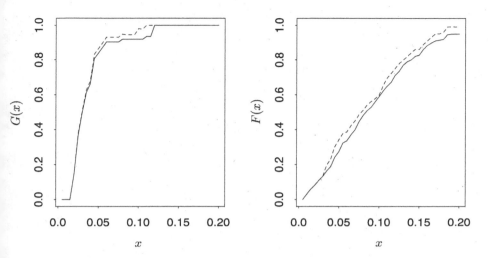

FIGURE 4.11
Estimates of $F(x)$ (left-hand panel) and of $G(x)$ (right-hand panel) for the redwood data. —— : $\hat{G}(\cdot), \hat{F}(\cdot)$; – – – : $\tilde{G}(\cdot), \tilde{F}(\cdot)$

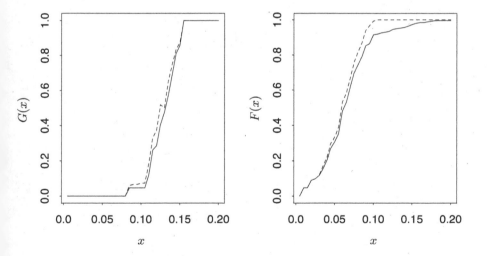

FIGURE 4.12
Estimates of $F(x)$ (left-hand panel) and of $G(x)$ (right-hand panel) for the cell data. —— : $\hat{G}(\cdot), \hat{F}(\cdot)$; – – – : $\tilde{G}(\cdot), \tilde{F}(\cdot)$

5

Nonparametric methods

CONTENTS

5.1   Introduction

In some studies, the questions of scientific interest can be addressed without specifying a parametric model for the data. A venerable, and generic, example is design-based inference for randomised experiments, which dates back at least to the early work of R. A. Fisher at Rothamsted; see, for example, Fisher, R.A. (1925, 1935).

In this chapter, we discuss methods for analysing spatial point pattern data that are not tied to particular parametric families of models. The material builds directly on the discussion of second-order properties in Section 4.6.

5.2   Estimating weighted integrals of the second-order intensity

Recall that one definition of the K-function of a stationary process is

$$K(t) = 2\pi\lambda^{-2} \int_0^t \lambda_2(s)s\,ds \qquad (5.1)$$

where λ and $\lambda_2(s)$ are the intensity and second-order intensity, respectively. It turns out that several non-parametric inference problems for point process data can be solved by estimating a weighted integral,

$$K_\phi(t) = 2\pi\lambda^{-2} \int_0^t \phi(s)\lambda_2(s)s\,ds, \qquad (5.2)$$

for suitably defined, problem-specific functions $\phi(s)$. Before considering a specific application, we give the general result, from Berman and Diggle (1989).

It is convenient to re-express (5.2) as

$$K_\phi(t) = 2\pi\lambda^{-2}H(t)$$

where

$$H(t) = \int_0^t \phi(s)\lambda_2(s)s\,ds. \qquad (5.3)$$

Using integration by parts to evaluate (5.3), and substituting from (5.1), we obtain

$$H(t) = \lambda^2(2\pi)^{-1}\left(K(t)\phi(t) - \int_0^t K(s)\phi'(s)ds\right) \qquad (5.4)$$

Estimation of $H(t)$ is now straightforward, because $\phi(\cdot)$ is a known function and we can substitute existing estimators for λ and for $K(t)$ into (5.4). In practice, the integration on the right hand side of (5.4) must be carried out numerically, but this is usually straightforward and numerically more stable than would have been the case for the direct numerical evaluation of (5.3). Given a sufficiently large data-set, it would also be feasible to substitute an estimate of $\lambda_2(t)$ into (5.3) and apply numerical integration to the right hand side of (5.3), but for the reasons discussed in Section 4.2, we prefer the estimator $\hat{H}(t)$ based on (5.4).

5.3   Nonparametric estimation of a spatially varying intensity

Suppose that the available data are a partial realisation of a Cox process, and that we wish to estimate the realisation of $\Lambda(x)$, the underlying intensity process. A simple and intuitively sensible estimator would consist of counting, for each location x, the number of events of the process within a distance h of x and scaling by πh^2, the area of a disc of radius h. In practice, we shall need to adjust this simple estimator to allow for edge-effects when x is close to the boundary of the study region, but we can ignore this complication for the time being. How should we choose the value of h? One way is to consider the mean square error of the resulting estimator.

Diggle (1985b) derived the mean square error of the estimator on the assumption that the underlying process is a stationary, isotropic Cox process. If the driving intensity of a Cox process has expectation μ and covariance function $\gamma(u)$, then the Cox process itself has intensity $\lambda = \mu$ and second-order intensity $\lambda_2(u) = \gamma(u) - \mu^2$. Let $N(x, h)$ denote the number of events of the Cox process within distance h of the point x. Then, temporarily ignoring edge-effects, the non-parametric estimator of the realised value of $\Lambda(x)$ described above can be written as

$$\tilde{\lambda}(x) = N(x, h)/(\pi h^2). \tag{5.5}$$

We now consider the mean square error of $\tilde{\lambda}(x)$,

$$MSE(h) = \mathrm{E}[\{\tilde{\lambda}(x) - \Lambda(x)\}^2],$$

where the expectation is with respect to the distribution of the Cox process, i.e. with respect both to $\Lambda(\cdot)$ and to the points of the process conditional on $\Lambda(\cdot)$. Stationarity implies that $MSE(h)$ does not depend on x. Taking $x = 0$ and using a standard conditioning argument, we have that

$$
\begin{aligned}
MSE(h) &= \mathrm{E}_\Lambda[\mathrm{E}_N[\{N/(\pi h^2) - \Lambda(0)\}^2]] \\
&= \mathrm{E}_\Lambda[\mathrm{Var}_N\{N/(\pi h^2)\} + \{\mathrm{E}_N[N/(\pi h^2)] - \Lambda(0)\}^2], \quad (5.6)
\end{aligned}
$$

where $N = N(0, h)$. Conditional on $\Lambda(\cdot)$, the count $N(0, h)$ follows a Poisson distribution with both mean and variance equal to $\int \Lambda(x)dx$, where the integration is over the disc with centre 0 and radius h. Hence, (5.6) becomes

$$
\begin{aligned}
MSE(h) &= \mathrm{E}_\Lambda\left[\int \Lambda(x)dx/(\pi h^2)^2 + \int\int \Lambda(x)\Lambda(y)dydx \right. \\
&\qquad\left. -2\int \Lambda(x)\Lambda(0)dx + \lambda(0)^2\right] \\
&= \lambda/(\pi h^2) + \int\int \lambda_2(||x - y||)dydx - 2\int \lambda_2(||x||)dx + \lambda_2(0).
\end{aligned}
$$

Now, use the fact that $\int \lambda_2(||x||)dx = \lambda^2 K(h)$ to give

$$MSE(h) = \lambda_2(0) + \lambda\{1 - 2\lambda K(h)\}/(\pi h^2) + (\pi h^2)^{-2}\int\int \lambda_2(||x - y||)dydx. \tag{5.7}$$

The first term on the right hand side of (5.7) does not depend on h, and it follows that the value of h which minimises $MSE(h)$ also minimises

$$M(h) = \lambda\{1 - 2\lambda K(h)\}/(\pi h^2) + (\pi h^2)^{-2}\int\int \lambda_2(||x - y||)dydx. \tag{5.8}$$

The first of the two terms on the right hand side of (5.8) is an explicit function of $K(h)$, and can be estimated by substituting the standard estimator $\hat{K}(\cdot)$.

The double integral in the second term can be converted to a single integral of the form (5.2) using polar coordinates, and can therefore be estimated as described in Section 5.1.

We now consider how to deal with edge-effects in the estimator $\tilde{\lambda}(x)$. Several methods of edge-correction are available. The one used by Diggle (1985b) in the one-dimensional case and extended to the two-dimensional case by Berman and Diggle (1989) replaces the denominator πh^2 in (5.5) by the area of intersection of the relevant disc with the study region. Hence, if the data are observed on a region A and $B(x, h)$ denotes the disc with centre x and radius h, the edge-corrected estimator is

$$\hat{\lambda}(x) = N(x, h)/|A \cap B(x, h)| \tag{5.9}$$

A further, and in this case largely cosmetic, refinement can be made by interpreting the estimator as a kernel estimator (Silverman, 1981). Define a kernel function $k(u)$ to be any radially symmetric, bivariate pdf (expressed in polar coordinates); thus, $k(u) \geq 0$ for all $u \geq 0$ and

$$2\pi \int_0^\infty k(u)u\,du = 1.$$

Then, $k_h(u) = h^{-2}k(u/h)$ is also a radially symmetric pdf for any $h > 0$, and a kernel estimator of a bivariate pdf $f(x)$, based on data $x_1, ..., x_n$, takes the form

$$\hat{f}(x) = n^{-1} \sum_{i=1}^n k_h(x - x_i).$$

The estimator $\tilde{\lambda}(x)$ can now be seen as a special case of the kernel estimator, with kernel function

$$k(u) = \begin{cases} (\pi u^2)^{-1} & : \quad 0 \leq u \leq 1 \\ 0 & : \quad u > 1. \end{cases} \tag{5.10}$$

Note, however, that the expressions given here for $\tilde{\lambda}(x)$ and $\hat{f}(x)$ differ by a factor of n, because the intensity is a mean number of events per unit area and, unlike the pdf, does not integrate to 1.

Viewed in this light, the edge-corrected estimator $\hat{\lambda}(x)$ can be written as

$$\hat{\lambda}(x) = \sum_{i=1}^n k_h(u) / \int_A k_h(||x||)dx$$

Most applications of kernel density estimation would use a smoother kernel function than (5.10), for example the quartic

$$k(u) = \begin{cases} 3\pi^{-1}(1 - u^2)^2 & : \quad 0 \leq u \leq 1 \\ 0 & : \quad u > 1. \end{cases} \tag{5.11}$$

In this case one way to choose the value of h is to calibrate (5.11) against (5.10)

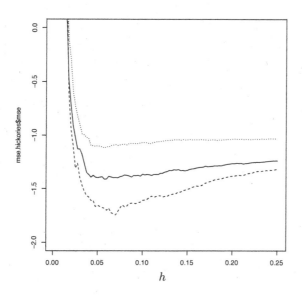

FIGURE 5.1

Estimates of the mean square error, $\hat{M}(h)$ of a non-parametric intensity estimator applied to each of the three major species groupings in Lansing Woods. $-\;-\;-\;-\;-\;-$: hickories; $-\;-\;-$: maples; .... : oaks.

by equating second moments. For the bivariate random variable, (X, Y) say, with pdf (5.11) the expectation of $U^2 = X^2 + Y^2$ is $h^2/4$, whereas with the circular uniform pdf (5.10) the corresponding expectation is $h^2/2$. Hence, if h_0 is the optimum bandwidth for a uniform kernel according to (5.8), the calibrated bandwidth for the quartic kernel is $h = h_0\sqrt{2}$.

In the original setting of non-parametric probability density estimation, Silverman (1981) points out that to obtain an estimate $\hat{f}(x)$ with good properties, the precise choice of kernel function is relatively unimportant by comparison with the choice of h. This is also generally true in the present context. An exception is when the underlying intensity surface is highly variable, in particular including large empty regions, and interest is in the ratio of intensities between two data-sets; this question arises naturally, for example in the application of point process methods to case-control studies in epidemiology, as we shall discuss in Chapter 9.

5.3.1 Estimating spatially varying intensities for the Lansing Woods data

Figure 5.1 shows estimates of the function $M(h)$ obtained by applying (5.8) to each of the three major species groupings of the Lansing Woods data shown

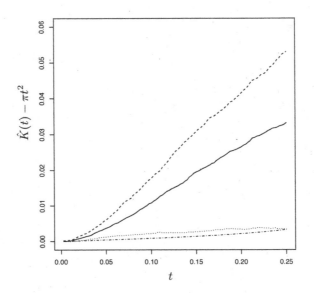

FIGURE 5.2
Estimates $\hat{K}(t) - \pi t^2$ for each of the three major species groupings in Lansing Woods. —— : hickories; – – – : maples; .... : oaks. Pointwise values of two standard errors assuming an underlying homogeneous Poisson process are shown as $-\cdot-\cdot-$

originally as Figure 2.11. We suggest that a plot of $\hat{M}(h)$ should be used as a guide to the choice of h, rather than as an automatic procedure. In this respect, we emphasise two particular features. Firstly, because $\hat{M}(h)$ is an empirical quantity and therefore subject to sampling variation, multiple local minima are to be expected in the neighbourhood of the optimal value of h where $M(h)$ is relatively flat. Secondly, and as illustrated here by the plot for the oaks, near-monotonicity of $\hat{M}(h)$ is a useful indication that there is little or no evidence in the data for spatial variation in intensity. When the underlying Cox process reduces to a homogeneous Poisson process, for which $\Lambda(x)$ is constant and $\gamma(u) = 0$ for all u, the theoretical form of $MSE(h)$ reduces to $MSE(h) = (\pi h^2)^{-1}$, which is monotone decreasing in h. The estimates shown in Figure 5.1 have been scaled so that they in fact estimate $\{MSE(h) + \lambda_2(0)\}/\lambda^2$. This implies that for a homogeneous Poisson process, $\hat{M}(h) =\to -1$ as $h \to \infty$.

The estimated K-function can also be used directly to confirm that the data are indeed spatially aggregated, and that a Cox process is a reasonable working model, before constructing an estimate of the realised intensity surface $\Lambda(x)$. Figure 5.2 shows $\hat{K}(t) - \pi t^2$ for each of the three species. The pointwise two-standard-error limits for a homogeneous Poisson process with the same intensity as the oaks are also shown as a dot-dashed line. This indicates that the second-order properties of the oaks are close to, albeit sig-

FIGURE 5.3

Kernel estimates of $\lambda(x)$ for the hickories (left-hand panel) and maples (right-hand panel) in Lansing Woods, using a quartic kernel with bandwidth $h = 0.1$.

nificantly different from, those of a homogeneous Poisson process whereas the hickories and maples are unequivocally spatially aggregated.

We therefore interpret Figure 5.1 as follows. For both the hickories and the maples, the minimum estimated mean square error is substantially less than -1, indicating spatial variation in intensity, whereas for the oaks the estimate $\hat{M}(h)$ is approximately constant for h greater than about 0.03 and never substantially less than -1. For the maples, there is a reasonably well-defined minimum of $\hat{M}(h)$ around $h \approx 0.07$, whereas the trace of $\hat{M}(h)$ for the hickories is minimised at $h \approx 0.06$, but is then rather flat until about $h \approx 0.09$. Our preliminary conclusion is that there is substantial spatial variation in the intensity of hickories and maples in Lansing Woods, whereas the oaks, although departing significantly from complete spatial randomness on the evidence of Figure 5.2, show approximately constant intensity. In general, we would consider using non-parametric estimation of spatially varying intensity $\lambda(x)$ only when, as is the case here, the data-set contains several hundred events or more, and when there is unequivocal evidence of substantial variation in $\lambda(x)$.

The two panels of Figure 5.3 show the estimates of $\lambda(x)$ obtained for the hickories and maples, using the quartic kernel (5.11) with $h = 0.07\sqrt{2} \approx 0.1$ in both cases to make the estimates directly comparable.

Notice that the two estimated surfaces $\hat{\lambda}(x)$ are essentially complementary to each other, in the sense that the intensity of hickories is high where the intensity of maples is low, and *vice versa*, suggesting that the two species occupy distinct ecological niches. We obtained a qualitatively similar result in Section (7.1.1) using a parametric model for $\lambda(x)$. However, with the selected bandwidth $h = 0.1$, the non-parametric estimates capture more of the variation in the data, the contour lines showing considerably more complicated behaviour than those of the parametric model. As noted earlier, the oaks display a pattern which is close to Poisson, implying in the present context a near-constant intensity, or an ability to exploit either of the two niches favoured by the hickories and by the maples. Of course, at a sufficiently small scale, we would expect to find competitive interactions between near-neighbouring trees, in violation of the Cox process assumption. Note, in this context, that the average area per tree in the whole forest (including a small number of miscellaneous trees not considered in this analysis) is approximately 35 square metres.

In contrast to the results for the parametric analysis of these data that we shall present in Section 8.2, if we assume intensity-reweighted stationarity and estimate $K_I(t)$ by using our non-parametric estimates of $\lambda(x)$ in Figure 5.3, then for both the hickories and the maples $\hat{K}_I(t)$ is significantly *less* than πt^2. Figure 5.4 shows, for each species, $\hat{K}_I(t)$ together with upper and lower envelopes from 19 simulations of an inhomogeneous Poisson process with intensity equal to the corresponding kernel estimate $\hat{\lambda}(x)$. The apparent regularity suggests small-scale inhibitory effects between neighbouring trees, but could also be a by-product of over-fitting a complicated surface $\hat{\lambda}(x)$.

The parametric analysis of these data that we shall discuss in Section 8.2.1 uses a log-quadratic model for the intensity, $\lambda(x)$. It could be argued that the nonparametric methodology described here is better able to describe subtle gradations in intensity over the study region. On the other hand, there is an attendant risk that a nonparametric smoothing method will find spurious features in the estimated intensity surface when used on sparse data. This underlines the importance of examining a plot of $\hat{M}(h)$ before computing the surface estimate $\hat{\lambda}(\cdot)$. A more important point relates to the theoretical requirement that $\lambda(x)$ should be bounded away from zero. In practice, this means that when the study-region A contain sub-regions where the intensity is close to zero, the estimator $\hat{K}_I(t)$ can be very sensitive to the choice of estimator for $\lambda(x)$.

5.4   Analysing replicated spatial point patterns

Nonparametric methods of inference are also appropriate when the data consist of replicate point patterns within a designed experiment, in which case inference can be based on the design rather than on an assumed stochastic model. For example, in Section 4.6 we considered the problem of estimating the sampling variance of an estimator for $K(t)$. There, we distinguished between situations in which the underlying process was or was not assumed to be a homogeneous Poisson process. In the second of these cases, we considered two different ways in which we might proceed, leading to model-based and design-based inference, respectively. For data consisting of a single point pattern, the design-based method used a form of pseudo-replication by divid-

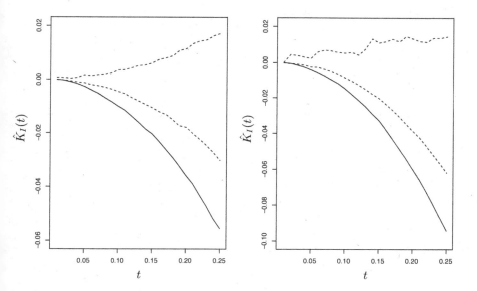

FIGURE 5.4
Estimates of $K_I(t)$ for the hickories (left-hand panel) and maples (right-hand panel) in Lansing Woods. $-----$: data; $---$ upper and lower envelopes from 19 simulations of the inhomogeneous Poisson process with intensity equal to the corresponding kernel estimate $\hat{\lambda}(\cdot)$.

ing the study region A into sub-regions. We now consider how to analyse data with genuine replication.

Genuine replication is obtained when an observed point pattern is the result of an experiment that can be repeated under identical conditions, thus producing a sequence of patterns which, by design, are exchangeable. It is then natural to analyse the resulting data using the design-based approach to inference.

As a motivating example, we consider the data shown in Figure 5.5. These consist of 12 point patterns, each of which identifies the locations of pyramidal neurons within a microscopic section of brain tissue taken from an area of the cingulate cortex (area 24, layer 2) of a human subject post-mortem (Diggle, Lange and Benes, 1991). The 12 subjects are presumed to be a random sample of normal brains, and it is of interest to quantify the typical spatial arrangement of pyramidal neurons, for later comparison with samples from abnormal subjects. It would be difficult to justify a stationary process model for the pattern presented by a single individual. Nevertheless, we can use the estimated K-function as a general summary measure of spatial aggregation, and the between-subject variation in estimated K-functions as the basis for inference.

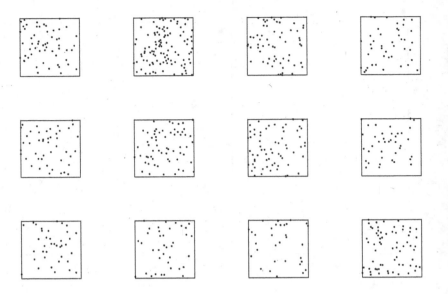

FIGURE 5.5
Locations of pyramidal neurons in brain tissue sections from 12 normal subjects.

5.4.1 Estimating the K-function from replicated data

Suppose that the data consist of r point patterns, each observed on a region A. Denote the ith of these by $\mathcal{X}_i = \{x_{i1}, ..., x_{in_i}\}$. When the patterns are strict replicates of an underlying process, the corresponding estimates of the K-function are identically distributed and a reasonable overall estimate can be obtained by simple averaging, as was done for the estimator $\tilde{K}(t)$ discussed briefly in Section (4.6.1). Because $K(t)$ is itself defined as a ratio, $K(t) = E(t)/\lambda$, a better strategy might be to pool separately estimates of λ and of $E(t) = \lambda K(t)$. This leads to $\hat{\lambda} = \sum n_i/(r|A|)$, which is the maximum likelihood estimator when the underlying process is a homogeneous Poisson process, and $\hat{E}(t) = \sum \hat{E}_i(t)/r$, where $\hat{E}_i(t)$ is the estimate of $E(t)$ obtained from the ith replicate. The resulting estimator for $K(t)$ is

$$\hat{K}(t) = \sum_{i=1}^{r} n_i \hat{K}_i(t) / \sum_{i=1}^{r} n_i, \qquad (5.12)$$

a weighted average of the individual estimates $\hat{K}_i(t)$.

The K-function is defined so as not to depend on the underlying intensity of events. We can therefore also construct a pooled estimated without assuming a common intensity across all replicates; this presumes that the hypothesis of a common K-function and varying intensity between replicates is scientifically plausible, as would be the case if the "replicates" were differentially thinned versions of a common underlying process. However, in this case we would again argue that the weighted average (5.12) is an appropriate estimator, since the dominant term in the variance of $\hat{K}_i(t)$ is of order n_i^{-1}. This also holds if the regions on which the separate patterns are observed differ in size. The distinction between assuming a common intensity λ or pattern-specific intensities λ_i might be expected to make a material difference. However, in the latter case, the argument leading to the estimator (5.12) holds directly, whereas in the former, it follows by using the estimators

$$\tilde{E}(t) = \sum_{i=1}^{r} |A_i| \hat{E}_i(t) / \sum_{i=1}^{r} |A_i|$$

and $\tilde{\lambda} = \sum_{i=1}^{r} n_i / \sum_{i=1}^{r} |A_i|_i$ together with the fact that $\hat{E}_i(t) = n_i \hat{K}_i(t)/|A_i|$.

For a design-based assessment of the sampling variance of $\hat{K}(t)$, we use a simple method based on the bootstrap (Efron and Tibshirani, 1993). Define *residual K-functions*,

$$R_i(t) = n_i^{0.5}\{\hat{K}_i(t) - \hat{K}(t)\} : i = 1, ..., r \qquad (5.13)$$

To a first approximation, the $R_i(t)$ are exchangeable under the assumption that the underlying processes may differ in their intensities, but are otherwise identical. We then construct a bootstrap sample of K-functions as

$$K_i^*(t) = \hat{K}(t) + n_i^{-0.5} R_i^*(t) : i = 1, ..., r,$$

where the $R_i^*(\cdot)$ are sampled at random with replacement from the set $\{R_1(\cdot), ..., R_r(\cdot)\}$, and compute a re-sampled $\hat{K}(t)$ as

$$\hat{K}^*(t) = \sum_{i=1}^{r} n_i \hat{K}_i^*(t) / \sum_{i=1}^{r} n_i.$$

Repeating this whole procedure, say s times, we then take the sample variance of the s values of $\hat{K}^*(t)$ as the bootstrap approximation to the sampling variance of $\hat{K}(t)$. Note that, because the whole of each residual K-function is re-sampled, the procedure gives a bootstrap estimate of the variance matrix of the vector of values of $\hat{K}(t)$ for the complete set of values of t under consideration, if required.

Figure 5.6 shows the estimate $\hat{K}(t) - \pi t^2$, as defined at (5.12), for the data in Figure 5.5, together with plus and minus two bootstrap standard error limits, computed from $s = 1000$ re-samples. Using a larger number of re-samples did not materially change the bootstrap variance estimates. All of the original estimated K-functions, and hence necessarily all of their bootstrap re-sampled counterparts, have $\hat{K}(t) = 0$ for small distances $t < 0.03$, approximately, indicating a small-scale inhibitory effect. At larger distances, the rapidly increasing width of the bootstrap standard error limits renders non-significant any further departure from an underlying $K(t) = \pi t^2$.

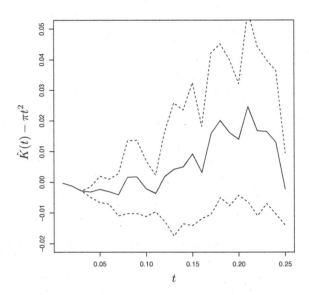

FIGURE 5.6
Pooled estimate $\hat{K}(t) - \pi t^2$ from 12 control subjects, with bootstrapped plus and minus two pointwise standard error limits

5.4.2   Between-group comparisons in designed experiments

We now extend our motivating example to include data from two additional experimental groups, consisting of post-mortem samples from subjects previously diagnosed as schizo-affective or schizophrenic. These data are shown in Figures 5.7 and 5.8. One of the patterns from the schizophrenic group has only two pyramidal neurons and we shall ignore it in the subsequent analysis.

Preliminary analysis of the counts n_i, using a log-linear Poisson regression model, suggested that the intensity of events varied significantly between groups. The observed mean numbers of cells per pattern in the three groups were 54.6 for the controls, 45.1 for the schizo-affectives and 37.4 (excluding the pattern with only two cells) for the schizophrenics. The likelihood ratio statistic to test for equality of the three underlying population means within the Poisson log-linear model was 33.3 on 2 degrees of freedom. There is no formal justification for assuming Poisson counts in this context, but unless the individual patterns are spatially aggregated, a Poisson approximation should be conservative (McCullagh and Nelder, 1989). Recall that the estimator $\hat{K}(t)$ remains valid when the underlying intensities vary between replicates, and therefore also when the intensities vary between groups. The two panels of Figure 8.9 show the resulting estimates of $K(t) - \pi t^2$ for the schizo-affectives and for the schizophrenics, together with pointwise limits constructed as plus and minus two bootstrap standard errors, and using a common vertical scale

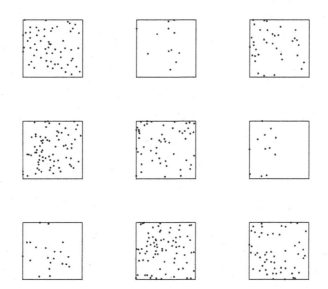

FIGURE 5.7
Locations of pyramidal neurons in brain tissue sections from 9 schizo-affective subjects.

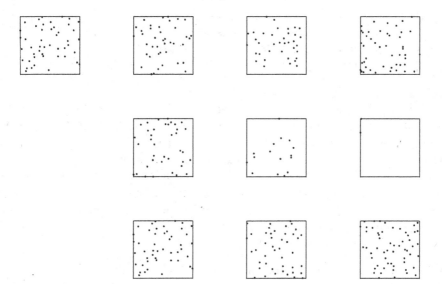

FIGURE 5.8
Locations of pyramidal neurons in brain tissue sections from 10 schizophrenic subjects.

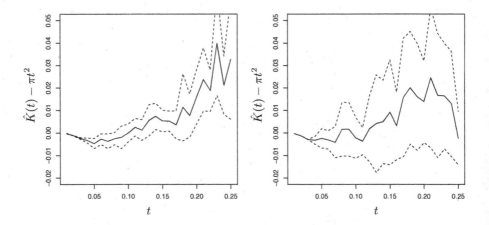

FIGURE 5.9
Pooled estimates $\hat{K}(t) - \pi t^2$ from 9 schizo-affective subjects (left-hand panel) and from 9 schizophrenic subjects (right-hand panel), with bootstrapped plus and minus two pointwise standard error limits.

for all three groups. Visual inspection of Figures 5.6 and 5.9 suggests that all three groups show small-scale inhibitory behaviour, as is to be expected in most micro-anatomical applications where the events represent reference locations for finite-sized objects. However, the groups appear to differ in their larger-scale behaviour; specifically, the schizo-affectives alone show apparently significant spatial aggregation. We now consider how to convert this visual assessment into a formal inference.

Let n_{ij} denote the number of events for the jth subject within the ith experimental group, and $\hat{K}_{ij}(t)$ the corresponding estimated K-function. Define $\hat{K}_i(t)$ to be the estimate (5.12) calculated from the r_i subjects in the ith group, and $\hat{K}_0(t)$ a weighted average of the $\hat{K}_i(t)$ with weights proportional to the total numbers of events, $n_i = \sum_{j=1}^{r_i} n_{ij}$, in the three groups. Now let $K_i(t)$ denote the expectation of $\hat{K}_{ij}(t)$ under repeated sampling. The null hypothesis is that this is the same in all three groups, hence $K_i(t) = K(t)$. Then, the $\hat{K}_i(t)$ estimate the corresponding $K_i(t)$, whilst $\hat{K}_0(t)$ estimates $K(t)$ under the null hypothesis that $K_i(t) = K(t)$ for all i.

To test the hypothesis that $K_i(t) = K(t)$, a statistic loosely analogous to the between treatment sum of squares in a classical analysis of variance is

$$BTSS = \sum_{i=1}^{3} n_i \int_0^{t_0} w(t)\{\hat{K}_i(t) - \hat{K}_0(t)\}^2 dt. \tag{5.14}$$

The null sampling distribution of T is intractable, but for a design-based inference we can again use a re-sampling method. For the jth subject within the ith group, define the *residual K-function*,

$$R_{ij}(t) = n_{ij}^{0.5}\{\hat{K}_{ij}(t) - \hat{K}_i(t)\} \tag{5.15}$$

The $R_{ij}(t)$ are, to a first approximation, exchangeable under the null or alternative hypotheses. Hence, if $R_{ij}^*(t) : j = 1, ..., r_i : i = 1, 2, 3$ are obtained from the original $R_{ij}(t)$ by re-sampling, we can construct a set of re-sampled K-functions under the null hypothesis as

$$\hat{K}_{ij}^*(t) = \hat{K}_0(t) + n_{ij}^{-0.5} R_{ij}^*(t) \tag{5.16}$$

For an approximate test, we then compare the observed value, $BTSS_1$ say, with values $BTSS_k : k = 2, ..., s$ from independent sets of re-sampled $K_{ij}^*(t)$.

We applied the bootstrap procedure with $t_0 = 0.25$, $w(t) = t^{-2}$ and 1000 re-samples. Note that in this context, each re-sample is a complete set of 30 estimated K-functions in three groups, but generated under the null hypothesis that the three underlying group-mean K-functions are equal. The resulting bootstrap p-value is 0.253, giving no reason formally to reject the null hypothesis.

Diggle, Lange and Benes (1991) used an unweighted version of the test statistic (5.14) in conjunction with a square root transformation of $\hat{K}(t)$. They also used random permutations of the residual K-functions, i.e. re-sampling without replacement, rather than the bootstrap re-sampling with replacement implemented here.

5.5   Parametric or nonparametric methods?

We have argued that for the application described in Section 5.4, parametric modelling assumptions would be hard to justify, and a nonparametric, design-based approach therefore seems natural. Of course, in this context an estimated K-function is just one of many possible *ad hoc* summary statistics that could be calculated from each pattern. It is a reasonable choice when there is scientific interest in the degree of spatial aggregation or regularity in the component patterns and in the extent to which this varies between groups of subjects. For more specific alternative hypotheses, other summary statistics may be preferable. Note also that Bell and Grunwald (2004) have analysed the same set of data using parametric methods. They fit pairwise interaction point process models using maximum pseudo-likelihood and accommodate differences between replicates by treating model parameters as random effects.

An immediate benefit of using likelihood-based methods when a suitable parametric model can be identified is that *ad hoc* methods of inference are no longer necessary. Also, information from independent replicates can be combined objectively, by adding the corresponding log-likelihood contributions. The corresponding cost is the reliance on additional assumptions, i.e. the correctness of the assumed model. An intermediate strategy is to use likelihood-based methods to estimate model parameters as summary statistics for each replicate, but to continue to use the randomisation distribution induced by the study design as the basis for inference. Diggle, Mateu and Clough (2000) report some empirical comparisons of the parametric and nonparametric approaches. Their results confirm, as expected, that the parametric approach is more powerful when its underlying assumptions are satisfied, but correspondingly less robust to departures from the assumptions.

6

Models

CONTENTS

6.1 Introduction

The basic building block for point process modelling is the Poisson process. As discussed in earlier chapters, the homogeneous Poisson process provides a benchmark of complete spatial randomness against which various kinds of pattern can be assessed. The essence of complete spatial randomness is the independence of the different events: knowing where some of the events are located does not help to predict where other events might be located. When the occurrence of an event at a particular location makes it more likely that other events will be located nearby, the resulting patterns display a kind of pattern which might loosely be described as *aggregated*. In contrast, when each event is likely to be surrounded by empty space, the overall pattern will be of a more *regular* spatial distribution of events.

In the remainder of this chapter, we will describe some simple constructions for these and other, more complex types of spatial pattern.

6.2   Contagious distributions

Historically, the first extension of the Poisson process as a model for spatial point patterns was advanced by Neyman (1939), who was concerned with possible models for the spatial distribution of insect larvae. Neyman postulated a Poisson process of egg-masses from which larvae hatch and subsequently move to positions relative to the corresponding egg-mass according to a bivariate distribution with pdf $h(\cdot)$. The probability that a larva from an egg-mass at $\mathbf{x}$ will subsequently be found within a region A is

$$P(\mathbf{x}; A) = \int_A h(\mathbf{y} - \mathbf{z})d\mathbf{y}.$$

Neyman then argued that without any knowledge of $h(\cdot)$ a model might reasonably be specified by a prescribed form for $P(\mathbf{x}; A)$. However, Skellam (1958) subsequently pointed out that the implied integral equation for $h(\cdot)$ may not be soluble; note that the required solution is a pdf which must not depend on A.

Rather more simply, suppose that "parent" events form a Poisson process with intensity ρ and that each parent, independently, produces a random number S of "offspring", all of which occupy the *same* position as their parent. The number of parents, M say, in a given region A therefore follows a Poisson distribution with mean $\rho|A|$. The number of offspring in A, $N(A)$ say, is $S_1 + \cdots + S_M$, and if the probability generating function (pgf) of S is $\pi_s(z)$, then the pgf of $N(A)$ is

$$\pi(z; A) = \exp[-\rho|A|\{1 - \pi_s(z)\}]. \tag{6.1}$$

Equation (6.1) defines the class of *generalized* Poisson distributions (Feller, 1968, Chapter 12, but note the difference in terminology). In the present context, such distributions are usually called *contagious*, following Neyman (1939). Neyman's Type A distribution is obtained by setting $\pi_s(z) = \exp\{-\mu(1 - z)\}$ and therefore corresponds to a non-orderly process of randomly distributed *point clusters*. A variation due to Thomas (1949) is to include parents in the final pattern. This avoids "clusters" of zero size and corresponds to $\pi_s(z) = z \exp\{-\mu(1 - z)\}$ Finally, if S has a logarithmic series distribution with $\pi_s(z) = 1 - \log\{1 + \beta(1 - z)\}/\log(1 + \beta)$ for some $\beta > 0$

then Y has a negative binomial distribution. The absence of any genuinely spatial clustering mechanism in the formulation of these contagious distributions should be noted. Furthermore, it is well known that the negative binomial distribution in particular can be derived also as a *compounded*, or mixed, Poisson distribution in which the parameter of a Poisson distribution is determined by random sampling from a gamma distribution.

Contagious distributions have long been fitted to quadrat count data, with apparent success. Evans (1953) gives a number of ecological examples, whilst Douglas (1979) contains a detailed description of the relevant statistical methodology. This approach provides at best an empirical description of *pattern*, and specific inferences about the underlying *process* should be avoided. In particular, it is far from clear that a negative binomial quadrat count distribution can be compatible with any spatial point process other than a point cluster process of the type defined by (6.1) or a non-ergodic process in which each complete realisation is a Poisson process, but whose intensity λ varies between realisations according to a gamma distribution; see, for example, the discussion of Matérn (1971).

The existence question was pursued by Diggle and Milne (1983a), who tried and failed to find a construction for which the resulting point process was stationary, ergodic, orderly and with negative binomial quadrat count distributions. They conjectured that no such process exists, and this was subsequently confirmed in unpublished work by Bob Griffiths (Department of Statistics, University of Oxford).

6.3   Poisson cluster processes

Poisson cluster processes, introduced by Neyman and Scott (1958), incorporate an explicit form of spatial clustering, and therefore provide a more satisfactory basis for modelling aggregated spatial point patterns. Their definition incorporates the following three postulates.

PCP1 Parent events form a Poisson process with intensity ρ.

PCP2 Each parent produces a random number S of offspring, realized independently and identically for each parent according to a probability distribution $p_s : s = 0, 1, \ldots$.

PCP3 The positions of the offspring relative to their parents are independently and identically distributed according to a bivariate pdf $h(\cdot)$.

Conventionally, and in the sequel unless explicitly stated otherwise, the final pattern consists of the offspring only. Some authors adopt a less restrictive definition involving the superposition of independent realisations of an arbitrary process, translated by the points of a Poisson parent process.

Poisson cluster processes as defined here are stationary, with intensity $\lambda = \rho\mu$ where $\mu = \mathrm{E}[S]$. They are isotropic if PCP3 specifies a radially symmetric pdf $h(\cdot)$.

To express the second-order properties in terms of the three postulates PCP1 to PCP3, let

$$h_2(x) = \int h(x)h(x - y)dx$$

be the pdf of the vector difference between the positions of two offspring from the same parent, and $H_2(\cdot)$ the corresponding cumulative distribution function. If we now consider an arbitrary event within a cluster of size S, the expected number of other events from the same cluster within a distance t is $(S - 1)H_2(t)$. The probability distribution of the size of the cluster to which an arbitrary event belongs is obtained by length-biased sampling from the cluster size distribution $p(s)$, hence $p^*(s) = sp(s)/\mu : s = 1, \ldots$. Averaging over the distribution $p^*(\cdot)$ then gives the expected number of related events within distance t of an arbitrary event as $\mathrm{E}[S(S - 1)]H_2(t)/\mu$.

Now, consider the expected number of unrelated events, meaning events from different clusters, within distance t of an arbitrary event. PCP1 implies that all such events are located independently of the original event, hence their expected number is just $\lambda\pi t^2$. Summing the contributions from related and unrelated events then gives

$$\lambda K(t) = \lambda\pi t^2 + \mathrm{E}[S(S - 1)]H_2(t)/\mu.$$

Finally, dividing by $\lambda = \rho\mu$ we obtain

$$K(t) = \pi t^2 + \mathrm{E}[S(S - 1)]H_2(t)/(\rho\mu^2). \tag{6.2}$$

Differentiation of (6.2), in conjunction with (4.3), then gives

$$\lambda_2(t) = \lambda^2 + \rho\mathrm{E}[S(S - 1)]h_2(t). \tag{6.3}$$

Note that the second term on the right hand side of (6.2) is non-negative, and monotone non-decreasing, and that $K(t) - \pi t^2$ approaches a constant, $c = \mathrm{E}[S(S-1)]/(\rho\mu^2)$, as $t \to \infty$. If S follows a Poisson distribution, $c = \rho^{-1}$. These results suggest a useful way of identifying whether a Poisson cluster process might be a reasonable model for an observed pattern, and if so a means of obtaining preliminary parameter estimates.

The variance of the quadrat count distribution for a Poisson cluster process is similarly obtained from (4.6) and (6.3) as

$$\mathrm{Var}\{N(A)\} = \rho\mu|A| + \rho E[S(S - 1)] \int_A \int_A h_2(x - y)dxdy.$$

General expressions for the nearest neighbour distributions of an isotropic Poisson cluster process are also available (Bartlett, 1975, Chapter 1). In the isotropic case, let $q(x, y)$ denote the probability that there are no offspring within a distance x of the origin, from a parent a distance y from the origin. Then, the distribution function of the point to nearest event distance is

$$F(x) = 1 - \exp\left(-2\pi\rho\int_0^\infty \{1 - q(x,y)\}y\,dy\right).$$

Because parents' locations are mutually independent, the distribution function of nearest neighbour distance follows immediately as

$$G(y) = 1 - \{1 - F(y)\}q^*(y),$$

where $q^*(y)$ denotes the probability that no offspring from the same parent as an arbitrary offspring, O say, lie within a distance y of O. In principle, the probabilities $q(x,y)$ and $q^*(y)$ are expressible in terms of the distributions specified in PCP2 and PCP3. Whilst the general expressions are not particularly illuminating, explicit results are obtainable in special cases; see, for example, Warren (1971) and Diggle (1975, 1978). Note also that $q^*(y)$ is an example of Van Lieshout and Baddeley's (1996) J-function.

In simulating Poisson cluster processes on a rectangular region, say $A = (0,a) \times (0,b)$, a useful device to avoid edge-effects is to impose periodic boundary conditions. Parents in A are first generated as a partial realisation of the appropriate Poisson process, as described in Section 4.4. Offspring are now attached to parents according to PCP2 and located according to PCP3, with the following exceptions:

1. Any generated x-coordinate of the form $ka + x$, for non-zero integer k and $0 < x < a$, is transformed to x.

2. Any generated y-coordinate of the form $kb + y$, for non-zero integer k and $0 < y < b$, is transformed to y.

In effect, the rectangle is converted to a torus by identifying opposite edges. When PCP2 specifies a Poisson distribution for the number of offspring per parent, the process can if required be simulated conditional on the total number of events in A by randomly allocating the events amongst the parents. Conditioning on the number of parents in A is straightforward.

Figure 6.1 shows parallel realisations of two Poisson cluster processes. Both have 25 parents on the unit square and an average of four offspring per parent. The position of each offspring relative to its parent follows a radially symmetric Gaussian distribution with pdf

$$h(x_1, x_2) = (2\pi\sigma^2)^{-1}\exp\{-(x_1^2 + x_2^2)/2\sigma^2\},$$

and $\sigma = 0.025$. In the left-hand panel, each parent has exactly four offspring. Notice that because formally distinct clusters coalesce it is difficult to identify the 25 sets of four offspring with any confidence. In the right-hand panel, the 100 offspring are randomly allocated amongst the 25 parents. Both patterns have used the same locations for parents, but the additional random element

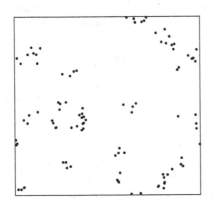 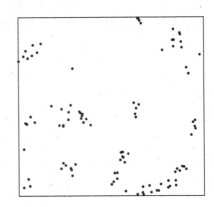

FIGURE 6.1
Realisations of two Poisson cluster processes, each with 25 parents on the unit square an average of four offspring per parent and radially symmetric Gaussian dispersion of offspring, with parameter $\sigma = 0.025$. In the left-hand panel, each parent has exactly four offspring. In the right-hand panel, offspring are randomly allocated amongst the 25 parents.

in the right-hand panel makes it still more difficult to identify the underlying process by visual inspection.

Poisson cluster processes can be extended to "multi-generation" processes in which the offspring become the parents of the next generation, and so on. This type of construction tends to be mathematically intractable, but it is intuitively appealing and we shall discuss it further in Section 5.5.

6.4   Inhomogeneous Poisson processes

A class of *non-stationary* point processes is obtained if the constant intensity λ of the Poisson process is replaced by a spatially varying intensity function, $\lambda(\mathbf{x})$. This defines the class of inhomogeneous Poisson processes, for which

IPP1 $N(A)$ has a Poisson distribution with mean $\int_A \lambda(x)dx$.

IPP2 Given $N(A) = n$, the n events in A form an independent random sample from the distribution on A with pdf proportional to $\lambda(x)$.

Figure 6.2 shows a partial realisation of an inhomogeneous Poisson process with A the unit square, $N(A) = 100$ and $\lambda(x_1, x_2) = \exp(-2x_1 - x_2)$. The

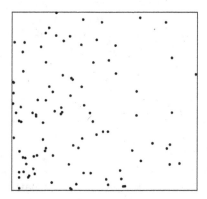

FIGURE 6.2
A realisation of an inhomogeneous Poisson process with 100 events on the unit
square and $\lambda(x) = \exp(-2x_1 - x_2)$

intensity gradient in the x_1-direction is immediately apparent, the gentler
gradient in the x_2-direction less so.

The inhomogeneous Poisson process provides a possible framework for the
introduction of covariates into the analysis of spatial point patterns via an
intensity function $\lambda(x) = \lambda\{z_1(x), z_2(x), ..., z_p(x)\}$. For example, suppose that
the locations of trees of a particular species are thought to follow a Poisson
process with intensity determined by height above sea-level, then a possible
model might be $\lambda(x) = \exp\{\alpha + \beta z(x)\}$, where $z(x)$ denotes height above
sea-level at the location x. Cox (1972a) refers to this as a "modulated Poisson
process".

Another example, which we will consider in more detail in Chapter 9,
would be a model for the point process of cases of a respiratory disease in
the vicinity of a point source of environmental pollution. In this example,
we might assume that case locations form an inhomogeneous Poisson process
with intensity $\lambda(x)$ given by

$$\lambda(x) = \lambda_0(x) f\{||x - x_0||, \theta\},$$

where $\lambda_0(x)$ corresponds to spatial variation in population density, x_0 is the
location of the point source, and $f(u, \theta)$ describes how the impact of the source
varies with distance, u.

Provided that $\lambda(x)$ is bounded away from zero, inhomogeneous Poisson
processes are re-weighted second-order stationary in the sense of Baddeley,
Møller and Waagepetersen (2000), with pair correlation function $\rho(t) = 1$ and
inhomogeneous K-function $K_I(t) = \pi t^2$.

The obvious method of simulating an inhomogeneous Poisson process is
via IPP2, whether with fixed or randomly generated $N(A)$. In special cases,

one-off algorithms can be devised. For the general case, Lewis and Shedler (1979) suggest an algorithm based on rejection sampling. In its simplest form, this consists of simulating a Poisson process on A with intensity λ_0 equal to the maximum value of $\lambda(x)$ within A, and retaining an event at $\mathbf{x}$ with probability $\lambda(x)/\lambda_0$.

6.5   Cox processes

The rationale behind the use of Poisson cluster processes as models for biological processes is that aggregated spatial point patterns might be generated by the clustering of groups of related events, as in Neyman's (1939) seminal paper. A second possible source of aggregation is environmental heterogeneity. Specifically, an inhomogeneous Poisson process with intensity function $\lambda(x)$ will produce apparent clusters of events in regions of relatively high intensity. The source of such environmental heterogeneity might itself be stochastic in nature. This suggests investigation of a class of "doubly stochastic" processes formed as inhomogeneous Poisson processes with stochastic intensity functions. Such processes are called Cox processes, following their introduction in one temporal dimension by Cox (1955). Explicitly, a spatial Cox process can be defined by the following postulates.

CP1 $\{\Lambda(x)\} : x \in \mathbb{R}^2\}$ is a non-negative-valued stochastic process.

CP2 Conditional on $\{\Lambda(x) = \lambda(x) : x \in \mathbb{R}^2\}$, the event form an inhomogeneous Poisson with intensity function $\lambda(x)$.

A Cox process is stationary if and only if its intensity process $\Lambda(x)$ is stationary, and similarly for isotropy. A convenient and expressive terminology is to refer to the Cox process "driven by" $\{\Lambda(x)\}$.

First-order and second-order properties are obtained from those of the inhomogeneous Poisson process by taking expectations with respect to $\{\Lambda(x)\}$. Thus, in the stationary case, the intensity is

$$\lambda = \mathrm{E}[\Lambda(x)].$$

Also, the conditional intensity of a pair of events at x and y, given $\{\Lambda(x)\}$, is $\Lambda(x)\Lambda(y)$, so that

$$\lambda_2(x, y) = \mathrm{E}[\Lambda(x)\Lambda(y)].$$

In the stationary, isotropic case this can be written as

$$\lambda_2(t) = \lambda^2 + \gamma(t), \tag{6.4}$$

where

$$\gamma(t) = \mathrm{Cov}\{\Lambda(x), \Lambda(y)\}$$

and $t = ||x - y||$. Note that, consistent with the notation introduced in Section 4.2, the covariance function $\gamma(t)$ of the intensity process is also the covariance density of the point process.

General expressions for $K(t)$ and $\mathrm{Var}\{N(A)\}$ then follow as in (4.3) and (4.6). Note that for the typical case in which $\gamma(t)$ takes non-negative values only, (6.4) is qualitatively similar to the corresponding expression (6.3) for a Poisson cluster process. More than this, processes in the two different classes can be shown to be equivalent. To see this, let $h(\cdot)$ be a bivariate pdf and construct an intensity process $\{\Lambda(x)\}$ by defining

$$\Lambda(x) = \mu \sum_{i=1}^{\infty} h(x - X_i) \tag{6.5}$$

for some $\mu > 0$, where the X_i are the points of a Poisson process. The Cox process driven by (6.5) is also a Poisson cluster process in which PCP2 specifies a Poisson distribution with mean μ, and PCP3 specifies the pdf $h(\cdot)$. Intuitively, this is because a Poisson distribution for the number of offspring per parent corresponds to the random allocation of offspring amongst parents. The equivalence of the two processes is established formally in Bartlett (1964).

By way of illustration, Figure 6.3 shows a realisation of a process of this type previously introduced in Section 5.3; we take $\mu = 4$ and a radially symmetric Gaussian $h(\cdot)$ with $\sigma = 0.05$. The process has again been conditioned to generate 100 events in the unit square and the realisation parallels the one shown in the right-hand panel Figure 6.1, to which it is identical in every respect save for the increased value of σ. The larger value of σ produces a more diffuse form of aggregation which, were it to be observed in the field, might suggest environmental heterogeneity rather than clustering. Whether or not such an interpretation were sound would then depend on further, biological investigation.

From a statistical viewpoint, the distinction between clustering and heterogeneity can only be sustained if additional information is available, for example in the form of covariates. Note that if we were able to model the intensity surface $\Lambda(x)$ through a regression equation in measured covariates, rather than as a realisation of a stochastic process, the resulting point process model would become an inhomogeneous Poisson process.

Matérn (1971) notes the difficulty of obtaining explicit expressions for the nearest neighbour distributions of a general Cox process. Conditional on the realisation $\lambda(x)$ of the intensity process, the probability that there are no events within a distance t of the origin is

$$\exp\left(-\int \lambda(x)dx\right), \tag{6.6}$$

where the region of integration is the disc with centre the origin and radius t. In principle, the distribution function of the distance from an arbitrary point to the nearest event is obtained by taking the expectation of (6.6) with

FIGURE 6.3
A realisation of a Cox process; see text for detailed explanation.

respect to the (infinite-dimensional) distribution of the surface $\Lambda(x)$ over a disc of radius t. In general, this is not an attractive proposition.

Kingman (1977) has argued that Cox processes provide a natural framework within which to model the spatial pattern of a population of reproducing individuals. Let G_n denote a point process which determines the locations X_i of individuals in the nth generation and suppose that reproduction obeys the following rules:

1. The number of offspring of the parent individual at x_i is a Poisson random variable with mean $\mu_i = \mu_i(G_n)$.

2. The positions of offspring relative to their parents are independently distributed according to a bivariate distribution with pdf $h(\cdot)$.

Rule (2) above is identical to postulate PCP3 of a Poisson cluster process, whilst (1) is similar to PCP2 but allows the μ_i to depend on the configuration of parent individuals; for example, μ_i might be a function of the number of parents within some prescribed distance of x_i. The locations of the offspring define the process G_{n+1}, and so on. It follows that G_{n+1} is a Cox process with $\{\Lambda(x)\}$ defined by

$$\Lambda(x) = \sum_{i=1}^{\infty} \mu_i(G_n)h(x - x_i),$$

where the x_i are the events of the nth generation and might, for example, be determined according to the "multi-generation" prescription discussed briefly at the end of Section 5.3.

In principle, any Cox process can be simulated by first simulating $\{\Lambda(x)\}$ on the appropriate region and then using the rejection sampling method for inhomogeneous Poisson processes as described in Section 5.4. More efficient methods can be devised for particular types of Cox process. For example, Cox processes defined by (6.5) are more efficiently simulated as Poisson cluster processes.

6.6 Trans-Gaussian Cox processes

One relatively flexible and tractable construction for Cox processes that is not based on their duality with cluster processes is the class of log-Gaussian processes, introduced by Møller, Syversveen and Waagepetersen (1998). These are defined by taking $\Lambda(x) = \exp S(x)$, where $S(x)$ is a Gaussian process. Moment properties follow from known properties of log-Gaussian distributions; specifically, if U is Normally distributed with mean m and variance v, then for any positive integer k,

$$E[U^k] = \exp(km + 0.5k^2 v). \qquad (6.7)$$

Now suppose that $S(x)$ is stationary with mean μ, variance σ^2 and correlation function $r(u) = \mathrm{Corr}\{S(x), S(x - u)\}$. Then, the intensity of the Cox process follows by setting $k = 1$ in (6.7), hence $\lambda = E[\Lambda(x)] = \exp(\mu + 0.5\sigma^2)$. Similarly, $E[\Lambda(x)\Lambda(x - u)] = E[\exp\{S(x) + S(x - u)\}]$. But $S(x) + S(x - u)$ is itself Normally distributed with mean 2μ and variance $2\sigma^2\{1 + r(u)\}$, hence $E[\Lambda(x)\Lambda(x - u)] = \exp[\mu + \sigma^2\{1 + r(u)\}]$. The covariance density and pair correlation function of the Cox process follow as

$$\gamma(u) = E[\Lambda(x)\Lambda(x - u)] - \lambda^2 = \exp(2\mu + \sigma^2)[\exp\{\sigma^2 r(u)\} - 1] \qquad (6.8)$$

and

$$\rho(u) = 1 + \gamma(u)/\lambda^2 = \exp\{\sigma^2 r(u)\}. \qquad (6.9)$$

Figure 6.4 shows a realisation of a log-Gaussian process in which $S(x)$ has $\sigma^2 = 1$, $\mu \approx 4.8$ to give $\lambda = 200$, and $r(u) = \exp(-4u)$. Also shown is the corresponding realisation of $\Lambda(x) = \exp\{S(x)\}$. The asymmetric appearance of the intensity surface, with relatively sharp peaks and flatter troughs, is a characteristic feature of this class of processes.

Less extreme asymmetry could be obtained by using a different transformation, for example by defining $\Lambda(x) = S(x)^2$. In this case, it follows from the moment properties of the bivariate Normal distribution that the intensity of the Cox process is $\lambda = \mu^2 + \sigma^2$, the covariance density is

$$\gamma(u) = 2\sigma^4 r(u)^2 + 4\mu^2 \sigma^2 r(u), \qquad (6.10)$$

and the pair correlation function is

$$\rho(u) = \{2\sigma^4 r(u)^2 + 4\mu^2 \sigma^2 r(u) + 1\}/(\mu^2 + \sigma^2)^2. \qquad (6.11)$$

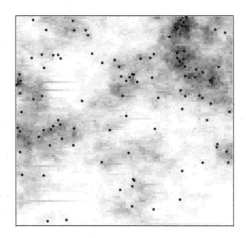

FIGURE 6.4
A realisation of a log-Gaussian Cox process (dots) superimposed on the realisation of the underlying intensity surface (grey-scale image); see text for detailed explanation.

6.7   Simple inhibition processes

The alternatives to the Poisson process described in Sections 5.3 and 5.5 share a tendency to produce aggregated patterns. Regular patterns arise most naturally by the imposition of a minimum permissible distance, δ say, between any two events. This may simply reflect the physical size of the biological entities whose locations define the point pattern (cf the discussion of Figure 1.3 in Section 1.1), or it may be a manifestation of more subtle effects such as competition between plants or territorial behaviour in animals. Processes of this sort which incorporate no further departure from complete spatial randomness are called simple inhibition processes, a notion which can be formalized in several non-equivalent ways. As a convenient piece of terminology, we define the *packing intensity* of a simple inhibition process as

$$\tau = \lambda \pi \delta^2 / 4,$$

where λ is the intensity. Thus, τ is the proportion of the plane covered by non-overlapping discs of diameter δ, or the expected proportion of coverage for a finite region A. Notice that the maximum possible packing intensity is attained by close-packed discs whose centres form an equilateral triangular lattice with spacing δ; thus $\tau \leq \tau_{max} = (\pi\sqrt{3})/6 \approx 0.907$.

Matérn (1960, Chapter 3) describes two types of simple inhibition process. In the first, a Poisson process of intensity ρ is thinned by the deletion of all pairs of events a distance less than δ apart. The probability that an arbitrary event survives is therefore $\exp(-\pi\rho\delta^2)$, and the intensity of the resulting inhibition process is

$$\lambda = \rho\exp(-\pi\rho\delta^2). \tag{6.12}$$

The corresponding packing intensity is at most $(4e)^{-1} \approx 0.092$, or about 10% of τ_{max}. The second-order properties are conveniently expressed by

$$\lambda_2(t) = \left\{ \begin{array}{ll} 0 & : \quad t < \delta \\ \rho^2\exp\{-\rho U_\delta(t)\} & : \quad t \geq \delta \end{array} \right. \tag{6.13}$$

where $U_\delta(t)$ denotes the area of the union of two discs each of radius δ and with centres a distance t apart. This follows because $\exp\{-\rho U_\delta(t)\}$ is the conditional probability that two events both survive, given that they are a distance $t \geq \delta$ apart.

In Matérn's second process, the events of a Poisson process are marked with times of birth and an event is removed if it lies within a distance δ of an "older" event. Expressions analogous to (6.12) and (6.13) can be obtained, but only by ignoring any consideration of whether or not the older event in question has itself previously been removed. Recognition of this last aspect leads to a simple sequential inhibition process, defined on any finite region A as follows. Consider a sequence of n events X_i in A. Then

SSI1 X_1 is uniformly distributed in A.

SSI2 Given $\{X_j = x_j, j = 1, ..., i - 1\}$, X_i is uniformly distributed on the intersection of A with $\{y : \|y - x_j\| \geq \delta, j = 1, ..., i - 1\}$.

Simple sequential inhibition is parameterized most naturally by its packing intensity, $\tau = n\pi\delta^2/(4A)$. Note that if too high a value of τ is prescribed, the sequential procedure may terminate prematurely. The maximum attainable packing intensity is a random variable whose distributional properties appear to be largely intractable; simulations by Tanemura (1979) suggest an expectation of about 0.547. Figure 6.5 shows the development of a realisation on the unit square with $\delta = 0.08$ and $n = 25, 50, 75$ and 100, the last corresponding to $\tau \approx 0.5$ The progressive development of regularity is clear.

Packing problems arise in many different contexts. Rogers (1964) gives a mathematical introduction to the subject. Bernal (1960) was among the first to use simple inhibition processes as models in the theory of liquids. Mannion (1964) discusses the "car-parking problem", which is simple sequential inhibition in one spatial dimension. Bartlett (1975, Chapter 3) relaxes the strict inhibition rule to replace the constant δ by a random variable, realized independently for each event, and uses the resulting process to model the spatial distribution of gulls' nests.

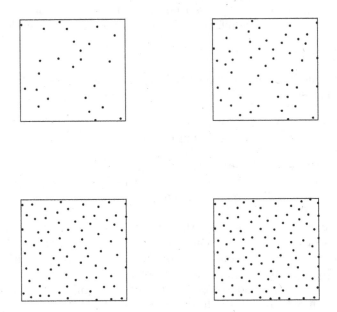

FIGURE 6.5
Progressive development of a realisation of a simple sequential inhibition process with $\delta = 0.08$ and $n = 25, 50, 75, 100$ events on the unit square.

6.8   Markov point processes

Many regular patterns require a more flexible description than can be provided by a strict, simple inhibition rule. For example, competitive interactions between plants may make it unlikely, but not impossible, that two individuals can survive in close proximity to each other.

For the simple inhibition processes described in Section 6.7 it is clear that the conditional intensity of an event at a point x, given the realisation of the process in the remainder of any planar region A, depends only on the existence or otherwise of an event within a distance δ of x. In other words, the process involves a form of local or Markovian dependence amongst events. For the reasons noted above, we might wish to preserve this local dependence, but introduce more flexibility into the model. This provides a motivation for the class of Markov point processes, introduced by Ripley and Kelly (1977).

Markov point processes are defined on an arbitrary, but fixed, finite region A. Each process is characterized by its likelihood ratio $f(\cdot)$ with respect to a Poisson process of unit intensity. Thus, if $\mathcal{X} = \{x_1, \ldots, x_n\}$ denotes any finite set of points in A, then $f(\mathcal{X})$ indicates in an intuitive sense how much more likely is the configuration of events $\mathcal{X}$ for the particular process than for a Poisson process of unit intensity. Usually, $f(\cdot)$ is defined up to a normalizing constant that cannot be calculated explicitly. Note that $f(\cdot)$ can always be factorized as a product of the form

$$f(\mathcal{X}) = \alpha \prod_{i=1}^{n} g_i(x_i) \prod_{j>i} g_{ij}(x_i, x_j) \ldots g_{12\ldots n}(x_1, x_2, \ldots, x_n). \tag{6.14}$$

We now define two points x and y in A to be *neighbours* if $||x - y|| < \rho$, where $\rho > 0$ is a prescribed value, called the *range* of the process. We further define a *clique* to be a set of mutual neighbours, and the *neighbourhood* of x to be the set of points $\{y \in A : 0 < ||x - y|| < \rho\}$. With these definitions, a point process is said to be *Markov of range ρ* if the conditional intensity at the point x, given the configuration of events in the remainder of A, depends only on the configuration in the neighbourhood of x. In this context, the conditional intensity is defined as the natural extension of the second-order conditional intensity as discussed in Section 4.2, except that the conditioning set is now an entire configuration of events in a specified region, rather than a single event at a specified location.

Ripley and Kelly (1977) establish the fundamental result that for a point process to be Markov of range ρ it is necessary that each g-function in (6.14) is identically unity unless its arguments constitute a clique; further conditions must be imposed, essentially to ensure that $f(\cdot)$ is integrable. There is a close link between this result for Markov point processes and the celebrated Hammersley-Clifford theorem for Markov random fields (Besag, 1974).

For a Poisson process of unit intensity, the likelihood of exactly n events in A at specified locations $x_1, \ldots, x_n$ is $\exp(-|A|)$, since $N(A)$ follows a Poisson distribution with mean $|A|$, the distribution of events given $N(A)$ is uniform on A and there are $n!$ equally likely permutations of $x_1, \ldots, x_n$. Thus, the likelihood function of a Markov point process can in principle be written as $f(\mathcal{X}) \exp(-|A|)$. In practice, this is of limited use because the normalizing constant is unknown.

Particular examples of Markov point processes include the Strauss process, for which

$$f(\mathcal{X}) = \alpha \beta^n \gamma^s \tag{6.15}$$

where n is the number of events in $\mathcal{X}$, s is the number of distinct pairs of neighbours, α is the normalizing constant, β reflects the intensity of the process and γ describes the interactions between neighbours. The case $\gamma = 1$ gives a Poisson process with intensity β, whilst $\gamma = 0$ gives a simple inhibition process, each realisation of which is a partial realisation of a Poisson process but conditioned by the requirement that no two events in the region A may be neighbours. This last process is formally different from simple sequential inhibition, but its statistical properties appear to be very similar (Ripley, 1977). Values of γ between 0 and 1 represent a form of non-strict inhibition. In the original paper, Strauss proposed (6.15) with $\gamma > 1$ as a model for clustering. Unfortunately, this results in an explosion of the process, with an infinite number of events in A. This can be seen intuitively from the form of (6.15), in which the exponent s can be of order n^2, whereas adjustment of the intensity via the parameter β can only absorb a term of order n. Kelly and Ripley (1976) establish this result formally.

If we fix n in (6.15), this results in a valid probability distribution for $\mathcal{X}$ for any non-negative γ. However, when $\gamma > 1$ the resulting patterns tend to exhibit an extreme form of clustering as the most likely configurations of events is one in which all n events from a single cluster of mutual neighbours. This is discussed in more detail in Gates and Westcott (1986).

6.8.1 Pairwise interaction point processes

The class of *pairwise interaction processes* is defined by

$$f(\mathcal{X}) = \alpha\beta^n \prod_{j\neq i} h\{||x_i - x_j||\}, \qquad (6.16)$$

where α and β are as in (6.15), $h(u)$ is non-negative for all u and the product is over all pairs of distinct points in $\mathcal{X}$ (Ripley, 1977). As the Strauss process is a special case of (6.16), it follows from the earlier discussion that some further restriction on $h(\cdot)$ is required in order to define a valid point process. A sufficient condition is that $h(\cdot)$ is bounded and $h(u) = 0$ for all u less than some $\delta > 0$; this automatically limits the number of events in any finite region A by imposing a minimum permissible distance δ between any two events. However, models which allow $h(u) > 1$ still tend to produce unrealistic models. In particular, the properties of any such process generating n events in a fixed, finite region of space depend strongly on the packing density, $\tau = n\pi\delta^2/(4|A|)$. At small packing densities, realisations of processes that allow $h(u) > 1$ tend to be extremely aggregated. As the value of τ increases, the small-scale regularity induced by the minimum distance constraint dominates the tendency towards aggregation induced by allowing $h(u) > 1$ at distances $u > \delta$ until, in the limit as τ approaches its maximum permissible value, the only feasible realisations of the process are approximate versions of a regular, close-packed lattice, irrespective of the nature of the interaction at larger distances.

To generate simulated realisations of a Markov point process the following procedure can be used. For illustrative convenience, we confine our attention to the class of pairwise interaction processes defined by (6.16), and condition each realisation to produce n events in A. Let $\mathcal{X} = \{x_1, \ldots, x_{n-1}\}$ be any set of $n-1$ events in A and consider the possible addition of an event at the point y. It follows from (6.16) that

$$f(\mathcal{X} \cup y) = \alpha\beta^n \prod_{i} \prod_{j\neq i} h\{||x_i - x_j||\} \prod_{i=1}^{n-1} h\{||x_i - y||\}$$

and

$$f(\mathcal{X} \cup y)/f(\mathcal{X}) = \beta \prod_{i=1}^{n-1} h\{||x_i - y||\}. \qquad (6.17)$$

Since $h(\cdot)$ is bounded, say $h(u) \le k$, then

$$p(y) = \left[\prod_{i=1}^{n-1} h\{||x_i - y||\} \right] / k^{n-1} \qquad (6.18)$$

is a probability. Consider any set of n points in A to constitute an initial realisation and delete one of the n points at random to produce the set $\mathcal{X}$ of (6.17). Now generate a point y distributed uniformly in A and accept y with probability $p(y)$, otherwise reject and repeat until one such point y is accepted. The process defined by alternative depletion and replacement of points according to the above prescription converges to the Markov point process defined by (6.16), but conditioned to produce n events in A; see Ripley, 1977, and Preston, 1977, for a formal justification. The number of depletion-replacement steps needed to achieve approximate equilibrium is unknown, but it is obviously sensible to use a feasible initial realisation. Ripley (1979b) gives a FORTRAN subroutine and suggests that $4n$ depletions and replacements are adequate in practice. This is an early example, and to my knowledge the first implementation in the statistical literature, of what would now be called an MCMC algorithm. A minor variation is to carry out the depletion-replacement steps in a systematic order, so that each "sweep" of n deletion-replacement steps results in the re-positioning of every event in the initial realisation. This gives a somewhat quicker convergence by reducing the statistical dependence between realisations from successive sweeps.

Figure 6.6 shows a realisation of each of two pairwise interaction processes conditioned to produce 100 events in the unit square, together with the corresponding interaction functions $h(\cdot)$. The first of these is a simple inhibition process with interaction function

$$h_1(u) = \begin{cases} 0 & : \quad u < 0.05, \\ 1 & : \quad u \ge 0.05. \end{cases}$$

whilst the second has

$$h_2(u) = \begin{cases} 0 & : \quad u < 0.05, \\ 20(u - 0.05) & : \quad 0.05 \le u < 0.1, \\ 1 & : \quad u \ge 0.1, \end{cases}$$

Each of these interaction functions imposes a minimum distance of 0.05 between events. The second additionally discourages pairs of events a distance less than 0.1 apart and thereby produces a more regular pattern of events. Both simulations were conditioned to generate exactly 100 events on the unit square and used four complete sweeps of Ripley's MCMC algorithm.

Figure 6.7 shows a realisation of each of another two pairwise interaction

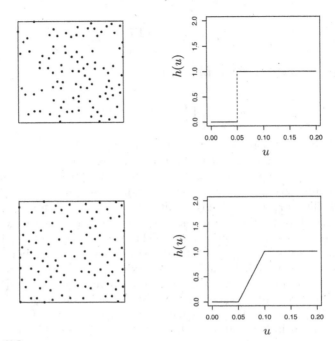

FIGURE 6.6
Realisations of two pairwise interaction point processes on the unit square and
their corresponding interaction functions (see text for detailed explanation).

processes, again conditioned to produce 100 events in the unit square, but
now with interaction function

$$h_3(u) = \begin{cases} 0 & : \quad u < \delta, \\ 2 - \delta/(u - \delta) & : \quad \delta \leq u < 2\delta, \\ 1 & : \quad u \geq 2\delta \end{cases} \qquad (6.19)$$

The upper row of Figure 6.7 corresponds to $\delta = 0.05$ in (6.19), with packing density $\tau \approx 0.2$. The resulting pattern is not grossly dissimilar to that
produced by a simple inhibitory process with $\delta = 0.05$ (cf the upper row of
Figure 6.6). The lower row of Figure 6.7 corresponds to $\delta = 0.01$, with packing
density $\tau \approx 0.008$. Now, the aggregative effect of allowing $h(u) > 1$ dominates,
to the extent that the simulated realisation consists of single, tight cluster of
events surrounded by empty space.

In most applications, it is more natural to consider the number of events as
a random variable because the study region, here the unit square, is itself only
part of a larger region on which the underlying process operates. However, for
inhibitory processes with interaction function $h(u) \leq 1$ for all u, whether we
treat the number of events in the study region as fixed or random does not
greatly affect the statistical properties of the resulting data. In the case of

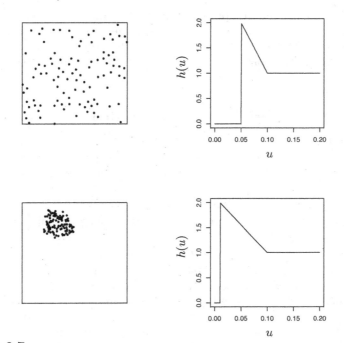

FIGURE 6.7
Realisations of two partly inhibitory, partly attractive pairwise interaction
point processes on the unit square and their interaction functions (see text for
detailed explanation).

partly attractive processes, for which $h(u) > 1$ for a range of values of u,
this is not necessarily so. As noted earlier, processes of this kind may not
even be well-defined as point processes on the plane, and when they are well-
defined they tend to generate extreme forms of clustering, as discussed in
Gates and Westcott (1986). In these circumstances, and as the lower row of
Figure 6.7 illustrates, if the fixed-number prescription is used to simulate a
pattern in a prescribed region A, then the partial realisations in sub-regions
of a given size and shape may have very different statistical properties than
the corresponding realisations on the whole of A. The author's conclusion is
that pairwise interaction processes should generally be used only as models for
regular spatial point patterns. However, the same conclusion does not apply
to spatio-temporal point patterns. In that setting, it is not only reasonable,
but often of direct interest, to model how the statistical properties of the
underlying process develop over time. From this point of view, Figure 6.5 can
be interpreted as a series of four snap-shots from a spatio-temporal simple
inhibition process. Spatio-temporal point processes are discussed in Chapters
10 to 13.

The MCMC algorithm described above can become computationally very expensive because the acceptance probability (6.18) can become extremely small. This is certainly the case when n is large and k, the designated upper bound on $h(u)$, is greater than one, but it can also apply to pairwise interaction functions that incorporate strong inhibitory interactions at relatively large distances.

6.8.2   More general forms of interaction

A more satisfactory construction for modelling aggregated patterns within the Markov point process framework is the area interaction process, as proposed by Baddeley and Van Lieshout (1995). In the simplest case, this process is defined by

$$f(\mathcal{X}) = \alpha\beta^n\gamma^{-A(\mathcal{X},\delta)}, \qquad (6.20)$$

where $A(\mathcal{X}, \delta)$ is the area of the union of discs, each of radius δ, centred on the points of $\mathcal{X}$.

When $\gamma = 1$ this reduces to a Poisson process with intensity β; when $\gamma < 1$ or $\gamma > 1$ its realisations exhibit spatial regularity or aggregation, respectively. In contrast to the Strauss process, as defined at (6.15), the exponent of the parameter γ in (6.20) is sub-linear in n, the number of points of $\mathcal{X}$, which prevents the process from exploding when $\gamma > 1$; instead, an area-interaction process with $\gamma > 1$ generates a stable form of spatial aggregation.

Baddeley and Møller (1989) consider a generalisation of the Markov point process construction in which the neighbourhood definition for a pair of events is allowed to vary dynamically according to the configuration of other events. One of their specific examples defines two events to be neighbours if their associated Dirichlet cells share a common boundary.

For a more detailed account of these and other constructions for Markov point processes, see Van Lieshout (2000).

6.9   Other constructions

6.9.1   Lattice-based processes

Various lattice-based processes can be used to generate regular spatial point patterns. These have limited appeal as models for natural phenomena, although the equilateral triangular lattice does represent an extreme of complete regularity and as such has been used as an idealized model of territorial behaviour within a population of mobile individuals (Maynard-Smith, 1974, Chapter 12).

A more pragmatic reason for studying lattice-based processes is their relative tractability by comparison with inhibition processes. In the early days of the development of spatial statistical methods, this tractability was exploited in the assessment of proposed inferential procedures, for example to make power comparisons amongst rival tests of CSR. In this context, deterministic lattice structures were used by Persson (1964) and Holgate (1965c), randomly thinned lattices by Brown and Holgate (1974), and the superposition of a deterministic lattice and a Poisson process by Diggle (1975).

6.9.2 Thinned processes

Many biological processes involve mortality, which in some instances is a reaction to an unfavourable environment. For example, the probability that a seedling survives may vary according to the amount of nutrient available in its immediate vicinity. Thinned point processes (Brown, 1979; Stoyan, 1979) provide a possible class of models for patterns which result from spatial variation in mortality.

A thinned point process is defined by a primary point process $N_0(dx)$ and a "thinning field" $Z(x)$, which is a stochastic process, independent of $N_0(\cdot)$, with realized values $0 \leq z(x) \leq 1$ for all x. Given realisations of $N_0(dx)$ and of $Z(x)$, the events x_i of $N_0(dx)$ are retained, independently, with respective probabilities $z(x_i)$. The corresponding realisation of the thinned point process $N(dx)$ consists of the retained events of $N_0(dx)$.

The second-order properties of $N(dx)$ are easily derived from those of $N_0(dx)$ and of $Z(x)$. In particular, in the stationary, isotropic case let μ and $\gamma(t)$ denote the mean and covariance function of $Z(x)$. Then, the second-order intensity function of $N(x)$ is

$$\lambda_2(t) = \lambda_{02}(t)\{\gamma(t) + \mu^2\}, \qquad (6.21)$$

where $\lambda_{02}(t)$ is the corresponding second-order intensity function of $N_0(dx)$. This follows because a pair of events of $N_0(dx)$ at locations x and y a distance t apart are both retained in the thinned process $N(dx)$ with probability $Z(x)Z(y)$. Using (4.3) and (6.21), and in an obvious notation, the K-functions of $N(dx)$ and $N_0(dx)$ are related by

$$K(t) = K_0(t) + \mu^{-2} \int_0^t \gamma(u) K_0'(u) du.$$

Note that if $\{N_0(d\mathbf{x})\}$ is a Poisson process, the thinned process is a Cox process. Thinned processes also provide one way of combining local interactions and stochastic environmental variation by taking $\{N_0(d\mathbf{x})\}$ to be a Markov point process. Figure 6.8 shows an example of a construction of this kind. The unthinned process $N_0(x)$ is a simple inhibitory process, with inhibition distance $\delta = 0.08$. To define the thinning field $Z(x)$, discs of radius 0.1 are centred on the events of a Poisson process of intensity 20. Then, $Z(x) = 1$ within the union of all such discs, $Z(x) = 0$ otherwise. The resulting thinned process $N(dx)$ displays small-scale regularity due to the inhibitory interactions, together with large-scale aggregation induced by the patches where $Z(x) = 1$.

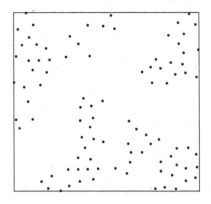

FIGURE 6.8
A thinning of an inhibitory pairwise interaction point process (see text for detailed explanation).

6.9.3   Superpositions

Another general construction to enrich the available range of models is to superimpose two or more component processes. For example, in Section 4.7 our analysis of the superposition of the two types of displaced amacrine cell was helpful in discriminating between the two competing scientific hypotheses concerning these data.

Provided that the component processes are independent, the second-order and nearest neighbour properties of the superposition are easily derived. In the bivariate case, let unsubscripted quantities refer to properties of the superposition, and subscripts 1 and 2 identify the corresponding properties for the component processes.

Note first that the component intensities λ_k add to give $\lambda = \lambda_1 + \lambda_2$. Now, using the result that $\lambda K(t)$ is the expected number of further events within distance t of an arbitrary event, together with independence of the component processes, we find that

$$\lambda K(t) = p\{\lambda_1 K_1(t) + \lambda_2 \pi t^2\} + (1-p)\{\lambda_2 K_2(t) + \lambda_1 \pi t^2\}$$

where $p = \lambda_1/\lambda$ is the probability that an arbitrary event is from component 1. It follows that

$$K(t) = \lambda^{-2}\{\lambda_1^2 K_1(t) + \lambda_2^2 K_2(t) + 2\lambda_1\lambda_2 \pi t^2\} \qquad (6.22)$$

Similarly, using $F(\cdot)$ and $G(\cdot)$ to denote point-to-event and event-to-event nearest neighbour distribution functions we find that

$$F(x) = 1 - \{1 - F_1(x)\}\{1 - F_2(x)\} \qquad (6.23)$$

and

$$G(x) = 1 - [p\{1 - G_1(x)\}\{1 - F_2(x)\} + (1-p)\{1 - G_2(x)\}\{1 - F_1(x)\}]. \quad (6.24)$$

This construction provides a convenient illustration of how second-order properties do not completely describe a point process. We consider the superposition of a homogeneous Poisson process and a Poisson cluster process, with respective K-functions $K_1(t) = \pi t^2$ and

$$K_2(t) = \pi t^2 + \rho^{-1} H(t) \quad (6.25)$$

where $H(\cdot)$ is the distribution function of the distance between two offspring from the same parent. Then, substitution of these expressions for $K_1(t)$ and $K_2(t)$ into (6.22) gives the K-function of the superposition as

$$K(t) = \pi t^2 + \lambda^{-2} \lambda_2^2 \rho^{-1} H(t). \quad (6.26)$$

Note that (6.26) is of the same form as (6.25), but with $\rho^* = \lambda^2 \rho / \lambda_2^2$ replacing ρ in (6.25), showing that the superposition of a Poisson process and a Poisson cluster process is indistinguishable from a pure Poisson cluster process on the basis of its second-order properties alone. The two *are* distinguishable by their different nearest neighbour properties. Setting $F_1(x) = G_1(x) = 1 - \exp(-\pi \lambda_1 x^2)$ in (6.23) and 6.24), we obtain the expressions

$$1 - F(x) = \exp(\pi \lambda_1 x^2)\{1 - F_2(x)\}$$

and

$$1 - G(x) = \exp(-\pi \lambda_1 x^2)[p\{1 - F_2(x)\} + (1-p)\{1 - G_2(x)\}].$$

The upper panels of Figure 6.9 show realisations of two such processes with identical intensities and K-functions but, as shown in the lower two panels of Figure 5.9, clearly different nearest neighbour properties. The difference is especially marked for the distribution function $F(\cdot)$, where the larger empty spaces apparent in the realisation of the pure cluster process by comparison with the superposition process translate into stochastically larger distances from sample points to nearest events.

6.9.4 Interactions in an inhomogeneous environment

Markov point processes are used to model interactions between events. Inhomogeneous Poisson processes are used to model environmental heterogeneity. A construction that combines these two features is obtained if we replace the constant β in the definition (6.16) of a pairwise interaction process by a function of position, $\beta(x)$. Then, the density for a configuration of events $\mathcal{X} = \{x_1, ..., x_n\}$ is given by

$$f(\mathcal{X}) = \alpha \prod_{i=1}^{n} \beta(x_i) \prod_{i \neq j} h(||x_i - x_j||).$$

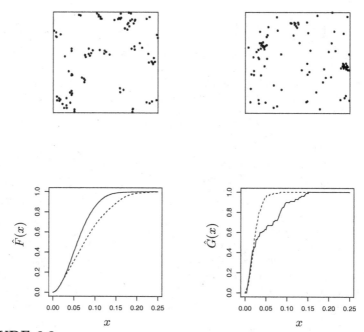

FIGURE 6.9

Realisations of two processes with identical second-order properties (upper panels) and their estimated nearest neighbour distributions functions (lower panels), with dashed and solid lines corresponding to a Poisson cluster process and the superposition of a Poisson cluster process and a homogeneous Poisson process.

Models of this kind are studied in Ogata and Tanemura (1986).

Figure 6.10 shows a realisation of a process with $\lambda(x_1, x_2)$ proportional to $\exp(-x_1 - 2x_2)$ and a simple inhibitory interaction function with minimum permissible distance $\delta = 0.03$ between any two events. The intensity gradient is clear. The small-scale inhibition is perhaps less obvious at first sight, but is made clear by comparing 6.10 with 6.2, which shows a realisation of an inhomogeneous Poisson process with the same intensity function.

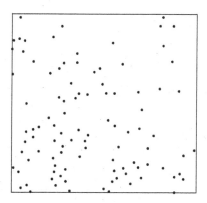

FIGURE 6.10
An inhomogeneous pairwise interaction point process (see text for detailed explanation).

6.10 Multivariate models

6.10.1 Marked point processes

One general construction for multivariate models is to label each of the events of a univariate point process by a categorical variable that distinguishes the different types of event. The categorical variable is called the *mark* variable, and the resulting process is an example of a *marked point process*.

In this context, and as discussed in Section 4.8, the simplest starting point for modelling is to assume that the marks of the different events are mutually independent and identically distributed. Under this *random labelling* hypothesis, all of the bivariate K-functions of the process are identical. More general models can be defined by allowing dependence amongst marks. For example, a *Markov random field* model for the marks would define, for each event x_i, the conditional distribution of the mark at x_i given the marks at all other events x_j. Markov random fields are very widely used in spatial statistics as models in their own right. An important early reference is Besag (1974); see also Rue and Held (2005).

6.10.2 Multivariate point processes

A different construction for multivariate models is as a set of possibly interdependent univariate point processes realised on the same space. As discussed in Section 4.5, if the separate univariate processes are stationary and independent, the bivariate K-function between any two component processes is equal to πt^2, whatever the marginal properties of the components.

6.10.3 How should multivariate models be formulated?

Although any multivariate model can be formally defined either as a marked point process or as a multivariate point process, in practice the choice between the two formulations will lead to different models. In particular, the "benchmark" hypotheses of random labelling and independence are different, except in the special case that the component processes are homogeneous Poisson processes, in which case random labelling and independence are equivalent. More generally, the range of potential models for multivariate point processes is so rich that there may be limited benefit in seeking to establish a comprehensive catalogue of "standard" models, as opposed to using the scientific context of a particular application to inform the modelling process.

To illustrate this point we can compare three hypothetical, but realistic, examples, each of which formally involves a bivariate point process or, equivalently from a formal theoretical perspective, a marked point process with binary-valued marks. We let P denote the unmarked point process and M the binary-valued mark process. The joint distribution of P and M can then be factorised in either of two equivalent ways, namely

$$[P, M] = [P][M|P] = [M][P|M]$$

where $[\cdot]$ is to be read as "the distribution of" and the vertical bar denotes conditioning.

Our first hypothetical example is in the area of human epidemiology. In this case, P identifies the locations of all members of a population at risk of contracting a particular disease, and M identifies which members of the population do, in fact, contract the disease. Here, the unmarked point process P is a physically sensible construct, and it would be natural to develop a model from the factorisation $[P, M] = [P][M|P]$. Furthermore, the marginal specification of $[P]$ is unlikely to be of scientific interest, and there would be no obvious value to the epidemiologist in devising an elaborate model for it. Hence, the focus of scientific interest is the conditional, $[M|P]$. In practice, the complete, unmarked process P is rarely observed. More commonly, data are obtained using a *case-control* study-design (Breslow and Day, 1980), consisting of all of the events of P with marks $M = 1$, and a random sub-set of the events with marks $M = 0$, called cases and controls, respectively. Point process methods for analysing data from case-control study-designs will be considered in Chapter 9.

Our second example concerns mineral exploration. Now, P identifies the locations of a number of exploratory drillings, and M identifies which drillings lead to the discovery of a commercially viable grade of the mineral in question. In contrast to the epidemiological example, the mark process M now derives from an underlying binary-valued random field $\{M(x) : x \in A\}$ which exists in it own right throughout the study region A, and which is the focus of scientific interest. Hence it would be natural in this context to begin by formulating a marginal model for M, within the factorisation $[P, M] = [M][P|M]$. This is what is done in the branch of spatial statistics known as geostatistics (Chiles and Delfiner, 1999, 2012; Diggle and Ribeiro, 2007), where typically the mark process is real-valued rather than binary. In that context, the joint distribution $[P, M]$ is rarely considered explicitly. On the contrary it is usually assumed implicitly that P and M are independent processes, i.e. that $[P, M] = [M][P]$. This would clearly be violated if exploratory drillings were deliberately sited at locations thought likely to yield commercially viable grades of ore. Schlather, Ribeiro and Diggle (2004) propose a number of tests of the hypothesis that $[M]$ and $[P]$ are independent processes. Diggle, Menezes and Su (2010) develop methodology for fitting a particular class of models for dependent $[M]$ and $[P]$, in which $[P]$ is a log-Gaussian Cox process with random intensity $\Lambda(x) = \exp\{S(x)\}$ and the mark of the ith event x_i is $M_i = S(x_i) + Z_i$, where the Z_i are mutually independent and Normally distributed. The scientific focus in Diggle, Menezes and Su (2010) is to use the marked point process data to make inferences about the spatially continuous process $S(x)$. Ho and Stoyan (2008) use a similar class of models but focus more on understanding the properties of the point process.

Our third, and in this case non-hypothetical, example concerns the joint distribution of nests of two species of ant, as considered by Harkness and Isham (1983), Hogmander and Sarkka (1999) and Baddeley and Turner (2000). In this case, neither factorisation of $[P, M]$ seems particularly helpful. Rather than attempting to model either a point process of ants of indeterminate species, or the species of a hypothetical ant at an arbitrary location, it would be more natural to model the two component processes, P_1 and P_2 say, in their own right, along with any possible interactions between ants of the same or different species. One way to do this is through multivariate extensions of Markov point processes, which we shall describe in Section 6.10.5.

We have argued that multivariate models should usually be related to the needs of specific applications. Nevertheless, it may be useful to give a few examples of specific multivariate constructions, and this we shall now do.

6.10.4   Cox processes

In a *multivariate Cox process*, the component processes are mutually independent Poisson processes, conditional on the corresponding intensities, $\lambda_j(x) : j = 1, ..., k$, which are formed as a realisation of a multivariate, non-negative valued stochastic process, $\Lambda(x) = \{\Lambda_1(x), ..., \Lambda_k(x)\}$. In what follows, we specifically discuss the bivariate case, $k = 2$.

Note firstly that any dependence structure between the two components of a bivariate Cox process arises only through dependence between the processes $\{\Lambda_1(x)\}$ and $\{\Lambda_2(x)\}$. In this sense, bivariate Cox processes provide a natural framework for modelling the joint reaction of two types of event to a stochastically heterogeneous environment, but do not incorporate any direct stochastic interactions amongst the events themselves.

Cox and Lewis (1972) adopt essentially the above definitions, but for point processes in time. Diggle and Milne (1983b) consider the extension of Cox and Lewis's work to bivariate spatial point processes, and give two examples that we shall describe later in this section.

As in the univariate case, nearest neighbour distributions are rarely tractable, but second-order properties can be expressed in terms of the corresponding properties of $\Lambda(x)$. For stationary $\Lambda(x)$, write

$$\lambda_j = E[\Lambda_j(x)] \tag{6.27}$$

and

$$\gamma_{ij}(u) = \text{Cov}\{\Lambda_i(x), \Lambda_j(y)\}, \tag{6.28}$$

where u is the distance between x and y. Then, consistent with earlier notation, λ_1 and λ_2 are also the intensities of the Cox process driven by $\{\Lambda(x)\}$. The second-order intensity functions are

$$\lambda_{ij}(u) = \gamma_{ij}(u) + \lambda_i\lambda_j \tag{6.29}$$

and (4.3) then gives

$$K_{ij}(t) = \pi t^2 + 2\pi(\lambda_i\lambda_j)^{-1} \int_0^t \gamma_{ij}(u)u\,du. \tag{6.30}$$

To provide an intuitively reasonable notion of "extreme positive correlation" within the class of bivariate Cox processes, Diggle and Milne (1983b) define a *linked* process as one for which

$$\Lambda_1(x) = \nu\Lambda_2(x),$$

for some positive constant $\nu = \lambda_1/\lambda_2$. Combining (6.27), (6.28) and (6.29) we deduce that $\lambda_{11}(u) = \nu\lambda_{12}(u) = \nu^2\lambda_{22}(u)$, with

$$\lambda_{12}(u) = \nu\{\gamma_{22}(u) + \lambda_{22}\}.$$

This shows that the covariance structure between the component point processes is simply a scaled version of the covariance structure *within* $\{\Lambda_2(x)\}$. Substitution into (6.30) then gives

$$K_{11}(t) = K_{22}(t) = K_{12}(t) = \pi t^2 + 2\pi\lambda_2^{-2} \int_0^t \gamma_{22}(u)u\,du.$$

To exemplify "extreme negative correlation" Diggle and Milne (1983b) defined a *balanced* bivariate Cox process as one for which $\lambda_1(x) + \Lambda_2(x) = \nu$, a positive constant. Note that in any such process, the superposition of the component processes is a homogeneous Poisson process.

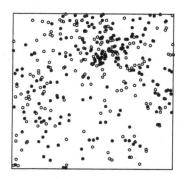 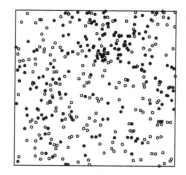

FIGURE 6.11

A linked (left-hand panel) and a balanced (right-hand panel) log-Gaussian Cox process (see text for detailed explanation).

Figure 6.11 shows realisations of a linked and a balanced log-Gaussian Cox process. In each case, the first component process has the same construction as was used for the univariate log-Gaussian Cox process illustrated in Figure 6.4. Also, both panels of Figure 6.11 use the same realisation for the first component. Visually, the positive dependence between the two components of the linked process is obvious, the negative dependence between the two components of the balanced process less so. This is essentially a consequence of the asymmetric nature of the underlying intensity process. Specifically, $\Lambda_1(x)$ consists of relatively sharp peaks on a relatively flat background plain. In the linked process, these peaks lead to easily discernible co-located concentrations of both types of events. In the balanced case, the plain of $\Lambda_1(x)$ becomes a high plateau of $\Lambda_2(x)$ with type 2 events scattered more or less randomly across it, whilst the peaks of $\Lambda_1(x)$ become sink-holes of $\Lambda_2(x)$ where the absence of type 2 events is not immediately obvious.

Linked and balanced Cox processes are extreme in the sense that the corresponding point-wise correlations between $\Lambda_1(x)$ and $\Lambda_2(x)$ are 1 and -1, respectively. Intermediate levels of correlation are most easily generated within the log-Gaussian framework, i.e. where $\Lambda_j(x) = \exp\{Z_j(x)\}$ and $\{Z_1(x), Z_2(x)\}$ is a bivariate Gaussian process. Constructing such processes is technically straightforward, finding examples that give flexible and intuitively appealing correlation structure less so; for a review, see Fanshawe and Diggle (2011). Also, the bivariate Gaussian is inflexible with regard to higher-order properties.

We emphasize that linking and balancing represent extreme positive second-order dependence only within the class of bivariate Cox processes. For example, the equality of all three K-functions in a linked bivariate Cox process reminds us that such processes are examples of random labelling mechanisms which, in other contexts, correspond to a form of non-association between the component processes (cf. the discussion of the displaced amacrine cell data in Section 4.7). It is also easy to construct non-Cox processes for which $K_{12}(t) > K_{jj}(t)$. For example, consider the following definition of a *linked pairs* bivariate Poisson process:

LP1 type 1 events form a homogeneous Poisson process of intensity λ;

LP2 each type 1 event has an associated type 2 event;

LP3 the positions of the type 2 events relative to their associated type 1 events are determined by a set of mutually independent realisations from a radially symmetric bivariate distribution.

Marginally, each component process is a homogeneous Poisson process, hence $K_{11}(t) = K_{22}(t) = \pi t^2$. However, if $H(t)$ is the distribution function of the distance between an event and its linked offspring, it is easy to show that $K_{12}(t) = \pi t^2 + \lambda^{-1} H(t)$.

Similarly, it is easy to construct non-Cox models that show more extreme negative second-order dependence than a balanced Cox process. One way to do this is by introducing inhibitory interactions between type 1 and type 2 events, as we now discuss.

6.10.5   Markov point processes

The formal definition of a Markov point process extends directly to multivariate processes, with essentially only notational changes. For example, a *bivariate pairwise interaction point process* requires the specification of three interaction functions, $h_{11}(\cdot)$, $h_{12}(\cdot)$ and $h_{22}(\cdot)$, rather than a single interaction function $h(\cdot)$.

We denote a bivariate configuration of points as $\{\mathcal{X}, \mathcal{Y}\}$, where $\mathcal{X} = \{x_1, ..., x_{n_1}\}$ and $\mathcal{Y} = \{y_1, ..., y_{n_2}\}$ are the configurations of type 1 and type 2 events, respectively. Then, in a bivariate pairwise interaction point process, the joint density of $(\mathcal{X}, \mathcal{Y})$ is

$$f(\mathcal{X}) = \alpha \beta_1^{n_1} \beta_2^{n_2} \prod_{i \neq j} h_{11}\{\|x_i - x_j\|\} \prod_{k \neq \ell} h_{22}\{\|y_k - y_\ell\|\} \prod_{p,q} h_{12}\{\|x_p - y_q\|\}$$

(6.31)

The component processes are *independent* if $h_{12}(u) = 0$ for all u, and *randomly labelled* if $h_{11}(u) = h_{12}(u) = h_{22}(u)$ for all u.

Even very simple forms of interaction function can lead to a wide range of bivariate behaviour by varying the model parameter values. For example, consider a simple inhibitory specification in which

$$h_{ij}(u) = \begin{cases} 0 & : \quad u \leq \delta_{ij} \\ 1 & : \quad u > \delta_{ij} \end{cases}.$$

If all three δ_{ij} are equal, the superposition of type 1 and type 2 events is

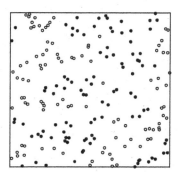 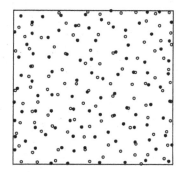

FIGURE 6.12
Two mutually inhibitory bivariate pairwise interaction point processes. The stronger inhibition is between events of opposite type in the left-hand panel, and between events of the same type in the right-hand panel (see text for detailed explanation).

a Strauss-type simple inhibition process, and the component processes are random thinnings of this. The component processes therefore exhibit a qualitatively similar, but less severe, form of regularity than does the superposition. If $\delta_{12} = 0$, the component processes are independent Strauss-type simple inhibition processes but there is no inhibition between events of opposite types and the superposition exhibits *less* regularity than do its components. Finally, if δ_{12} is large relative to δ_{11} and δ_{22} then the bivariate pattern will tend to produce segregated clumps of type 1 and type 2 events. Each component will then exhibit large-scale spatial aggregation but also, assuming that $\delta_{jj} > 0$, small-scale regularity within each segregated clump.

Figure 6.12 illustrates both cases. In the left-hand panel, $\delta_1 = \delta_2 = 0.007$ and $\delta_{12} = 0.07$, whilst in the right-hand panel, $\delta_1 = \delta_2 = 0.07$ and $\delta_{12} = 0.007$. Sarkka (1983), Baddeley and Møller (1989) and Hogmander and Sarkka (1999) discuss similar examples.

7

Model-fitting using summary descriptions

CONTENTS

7.1 Introduction

Classical methods of inference for spatial point processes are hampered by the intractability of the likelihood function for most models of interest. To some extent, this difficulty has been alleviated by the development of Monte Carlo methods for calculating approximate likelihoods, and we shall discuss the resulting methods of inference in the next chapter. Nevertheless, the more *ad hoc* methods of inference described in this chapter, which operate by comparing theoretical and empirical summary descriptions, remain useful for at least two reasons. Firstly, they are computationally straightforward, and consequently useful for rapid exploration of a range of possible models. Secondly, they provide direct, graphical methods for assessing model fit.

7.2   Parameter estimation using the K-function

In this section, we describe a method of parameter estimation using $K(t)$ and its estimator $\hat{K}(t)$. Analogous methods using the summary descriptions $G(y)$ or $F(x)$ could be implemented and might in some instances be necessary to allow identification of all the model parameters (cf. Section 5.8.3).

One attraction of an analysis using $\hat{K}(t)$ is that the mathematical form of $K(t)$ is known, either explicitly or as an integral, for a number of potentially useful classes of spatial point process. Plots of $\hat{K}(t)$ can therefore be used to suggest candidate models and to provide initial parameter estimates. To a limited extent this remains true in the case of less tractable models, for which $K(t)$ retains its tangible interpretation as an expectation. A glance back at Figure 6.1 should convince the reader that it can be extremely difficult to make reasonable guesses at parameter values merely by inspecting the data.

7.2.1   Least squares estimation

Suppose that our model incorporates a vector of parameters θ. Let $K(t; \theta)$ denote the theoretical K-function and $\hat{K}(t)$ the estimator (4.15) calculated from the data. A class of criteria to measure the discrepancy between model and data is given by

$$D(\theta) = \int_0^{t_0} w(t)[\{\hat{K}(t)\}^c - \{K(t; \theta)\}^c]^2 dt, \qquad (7.1)$$

where the constants t_0 and c, and the weighting function $w(t)$ are to be chosen. We then estimate θ to be the value $\hat{\theta}$ which minimizes $D(\theta)$.

An immediate question is how to choose t_0, c and $w(t)$. As noted earlier, there are good theoretical and practical reasons for not using too large a value of t_0. The power transformation c and the weighting function $w(t)$ present two opportunities to allow for the nature of the sampling fluctuations in $\hat{K}(t)$. These sampling fluctuations increase with t and so have a potentially wayward influence on the estimation of θ. By reducing the influence of large values of t we also make the precise choice of t_0 less critical. Besag (1977) pointed out that for a Poisson process the variance of $\sqrt{\hat{K}(t)}$ is approximately independent of t, so that $c = 0.5$ acts as a variance-stabilising transformation for patterns which are not grossly different from CSR. Empirical experience with real and simulated data, some of which is reported in the remainder of this chapter, suggests that $c = 0.5$ in conjunction with $w(t) = 1$ is a reasonable choice for fitting models to regular patterns, but that for aggregated patterns something more severe, say $c = 0.25$, is usually more effective. On similarly empirical grounds we would recommend that for data on a rectangular region with side-lengths a and b, t_0 should be no bigger than $0.25\min(a, b)$, but this choice can also to some extent be related to the model in question: taking t_0

small concentrates on small-scale effects and conversely. A reasonable practical strategy is to try a few different values of t_0 and of c in order to assess the sensitivity of the results to these choices.

A different strategy would be to fix $c = 1$ and use a weighting function inversely proportional to a guess at the sampling variance of $\hat{K}(t)$. For patterns that are close to Poisson, $w(t) = t^{-2}$ is a reasonable choice, although using this approach we need to be careful to avoid problems of numerical instability near $t = 0$.

The sampling distribution of θ can be assessed by repeated application to simulated realisations of the fitted model, although this may be a computationally expensive exercise for very large data-sets.

7.2.2   Simulated realisations of a Poisson cluster process

To provide an illustration for which the true value of θ is known, we use simulated data from a Poisson cluster process with Poisson numbers of offspring per parent and radially symmetric Gaussian dispersion of offspring relative to their parents. Using the additive property of two independent Gaussian random variables and a transformation to polar coordinates we deduce that for this model the distribution function of the distance between two offspring from the same parent is $H_2(t) = 1 - \exp\{-t^2/(4\sigma^2)\}$. Thus, using (6.2),

$$K(t) = \pi t^2 + \rho^{-1}[1 - \exp\{-t^2/(4\sigma^2)\}], \qquad (7.2)$$

where ρ is the mean number of parents per unit area and σ is the dispersion parameter of the radially symmetric Gaussian distribution. The resulting mean squared distance of an offspring from its parent is $2\sigma^2$. In a rather loose sense, the degree of aggregation in the model increases with a reduction in the value of either ρ or σ.

For various values of $\theta = (\rho, \sigma)$ we simulated 100 replicates and estimated θ from (7.1) using $t_0 = 0.05, 0.15, 0.25$ and $c = 0.125, 0.25, 0.5$. Figure 7.1 shows the empirical sampling distribution of $\log \hat{\rho}$ and $\log \hat{\sigma}$ when $(\rho, \sigma) = (100, 0.025)$, $t_0 = 0.125$ and $c = 0.25$, with each replicate consisting of 400 events on the unit square. The strong linear relationship with slope somewhere around -2 suggests that the product $\rho\sigma^2$ can be estimated with much higher precision than either ρ or σ separately. An explanation is that for small $t^2/(4\sigma^2)$, a series expansion of the exponential in (7.2) gives

$$K(t) \approx \pi t^2 + t^2/(4\rho\sigma^2)$$

so that, to this degree of approximation, ρ and σ^2 are not separately identifiable. The right-hand panel of Figure 6.1 showed a partial realization of this model, but with 100 rather than 400 events on the unit square and the values of ρ and σ re-scaled accordingly.

The simulation experiment confirmed that, at least for this model, using $c = 0.25$ or smaller makes the estimation procedure relatively insensitive to

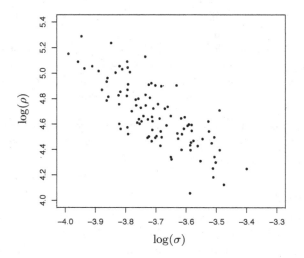

FIGURE 7.1
The empirical sampling distribution of $(\log \hat\rho, \log \hat\sigma)$ in a Poisson cluster process with 400 events in the unit square, $\rho = 100$ and $\sigma = 0.025$.

the choice of t_0. An exception to this general statement is that when σ is large relative to the dimensions of the study-region A, the model parameters are very poorly identified whatever value is chosen for c. In these circumstances the optimization algorithm occasionally failed to converge, or declared convergence to materially different estimates, depending on the starting value. The results reported here used a quasi-Newton method (Gill and Murray, 1972), as implemented in the R function optim().

7.2.3   Procedure when $K(t)$ is unknown

When $K(t; \theta)$ cannot be evaluated either explicitly or numerically, it can be replaced in (7.1) by $\bar{K}_s(t)$, the pointwise mean of estimates $\hat{K}_i(t)$ calculated from s simulated realizations of the model. If s is large, each evaluation of $D(\theta)$ may then be computationally expensive, in which case a sensible strategy is to start with a small value of s and use a robust minimization algorithm, such as the simplex algorithm of Nelder and Mead (1965), to find a first approximation to $\hat{\theta}$. The algorithm can then be restarted from this first approximation using a larger value of s and a more stringent stopping rule. In this context, Diggle and Gratton (1984) used a modification of the simplex algorithm in which s is automatically increased whenever the variation in $D(\theta)$ between different points θ in the simplex becomes comparable to the simulation-induced variation in repeated evaluation of $D(\theta)$ for fixed θ.

7.3 Goodness-of-fit assessment using nearest neighbour distributions

Any of the Monte Carlo tests of complete spatial randomness described in Chapter 2 can also be used to test the goodness of fit of a fully specified model, by simulating the appropriate model in place of CSR. Such tests are strictly invalid, and probably conservative, if parameters have been estimated from the data. However, this problem can be alleviated if we use a goodness-of-fit statistic which is only loosely related to the estimation procedure (cf. Section 1.7).

We shall therefore consider two goodness-of-fit statistics based on the EDF's of nearest neighbour and point to nearest event distances,

$$g_i = \int_0^\infty \{\hat{G}_i(y) - \bar{G}_i(y)\}^2 dy, \tag{7.3}$$

and

$$f_i = \int_0^\infty \{\hat{F}_i(x) - \bar{F}_i(x)\}^2 dx, \tag{7.4}$$

where $\hat{G}_1(y)$ is the EDF of nearest neighbour distance for the data, $\hat{G}_i(y), i = 2, \ldots, s$, are the EDF's from simulations of the model,

$$\bar{G}_i(y) = (s-1)^{-1} \sum_{j \neq i} \hat{G}_j(y),$$

and similarly for $\hat{F}_i(x)$ and $\bar{F}_i(x)$.

We carried out a simulation experiment to investigate the joint sampling distribution of the attained significance levels of Monte Carlo tests based on (7.3), (7.4) and on a third set of statistics,

$$k_i = \int_0^{0.25} \left[\{\hat{K}_i(t)\}^{\frac{1}{2}} - \{\bar{K}(t)\}^{\frac{1}{2}} \right]^2 dt, \tag{7.5}$$

which are related in an obvious way to the method of parameter estimation described in Section 6.1.

The simulation experiment involved generating data as a realization of a particular model, testing the goodness of fit of that model using Monte Carlo tests based on each of (7.3), (7.4) and (7.5) and repeating the entire procedure 100 times. This produced a 100×3 matrix of attained significance levels, which we examined by sample means, variances and correlations, and by scatter-plots. Each simulated data-set consisted of $n = 100$ events on the unit square. Three different models were used, to embrace complete spatial randomness, aggregation and regularity: (i) a homogeneous Poisson process; (ii) a Poisson cluster process, as in Section 6.1.2, with $(\rho, \sigma) = (25, 0.025)$; (iii) simple sequential inhibition, with $\delta = 0.08$. Realizations of these three models

TABLE 7.1
Estimated correlations amongst attained significance levels for three goodness-of-fit tests

Model	Estimated correlation between tests based on		
	g_i, f_i	g_i, k_i	f_i, k_i
Poisson process	0.088	0.047	0.215
Poisson cluster process	0.048	0.268	0.004
Simple sequential inhibition	−0.077	0.467	0.001

were shown in Figures 1.4, 5.2 and 5.6, respectively. In all cases, the sample means and variances were consistent with the theoretical values implied by a discrete uniform distribution of attained significance levels. The correlations amongst the results of the three tests are given in Table 6.1. The moderately large positive correlation between tests based on (7.3) and on (7.5) in the case of simple sequential inhibition is not surprising since the pattern of *small* inter-event distances is here of paramount importance. The remaining correlations are encouragingly small, and the scatter-plots showed no obvious non-linear relationships. In view of these results and remarks in Section 2.6, our preferred method of goodness-of-fit testing is to use both (7.3) and (7.4) in conjunction with the inequality (2.1). This amounts to accepting the model only if neither test indicates a significant lack of fit.

7.4 Examples

7.4.1 Redwood seedlings

In Figure 1.2 we showed the locations of 62 redwood seedlings in a square of side 23 metres approximately (data extracted by Ripley, 1977, from Strauss, 1975). The methods of preliminary testing described in Chapter 2 led to emphatic rejection of CSR for these data. Ripley (1977) reached the same conclusion, but did not suggest an alternative model. Strauss (1975) used a larger set of data to fit a pairwise interaction point process with interaction function

$$h(u) = \begin{cases} \gamma & : \quad u < \delta \\ 1 & : \quad u \geq \delta, \end{cases}$$

and $\gamma > 1$. Kelly and Ripley (1976) subsequently pointed out that, for the reasons given in Section 5.7, this model is only valid for $0 \leq \gamma \leq 1$, in which case it can only generate regular patterns. However, Strauss also noted that the apparent aggregation in these data is attributable to clustering of seedlings

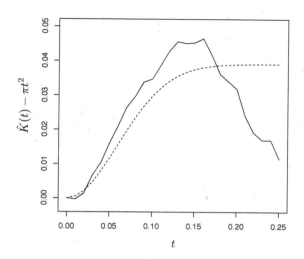

FIGURE 7.2

$K(t) - \pi t^2$ for the redwood seedlings. ——: data; − − −: fitted model.

around stumps which are known to be present in the plot but whose positions were not recorded. A Poisson cluster process is therefore plausible as a provisional model. We fit the particular model described in Section 5.3.2, in which the number of offspring per parent is Poisson and the dispersion of offspring relative to their parents follows a radially symmetric Normal distribution. Recall that for this model,

$$K(t) = \pi t^2 + \rho^{-1}[1 - \exp\{-t^2/(4\sigma^2)\}],$$

where ρ is the mean number of parents per unit area and $2\sigma^2$ the mean squared distance of an offspring from its parent.

Figure 7.2 shows a plot of $\hat{K}(t) - \pi t^2$ against t. Equating the maximum on the plot to the point $(4\sigma, \rho^{-1})$ gives initial parameter estimates $(\tilde{\rho}, \tilde{\sigma}) = (22.5, 0.040)$. Least squares estimation using (7.1) with $t_0 = 0.25$, $c = 0.25$ and $w(t) = 1$ gives estimates $(\hat{\rho}, \hat{\sigma}) = (25.6, 0.042)$. The behaviour of $\hat{K}(t)$ for $t > 0.16$ is superficially incompatible with the fitted $K(t)$, which is also shown in Figure 7.2. However, because the sampling fluctuations in $\hat{K}(t)$ increase with t, the fit proves statistically adequate. Goodness-of-fit tests based on (7.3) and (7.4) give nominal attained significance levels of 0.10 and 0.77 respectively or, in combination, $0.10 \leq p \leq 0.20$.

Graphical summaries of $\hat{G}(y)$ and $\hat{F}(x)$ for model and data are given in Figure 7.3. Note that the nearest neighbour EDF $\hat{G}(y)$ drifts briefly above the upper envelope from 99 simulations of the model, but that this does not in itself justify rejection of the model because it would imply a retrospective choice of test statistic. Figure 7.4 shows a realisation of the fitted model. Both

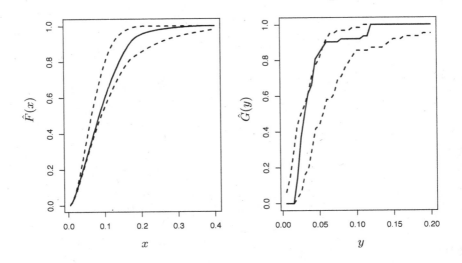

FIGURE 7.3
Goodness-of-fit of a Poisson cluster process to the redwood seedling data, using nearest neighbour (left-hand panel) and point-to-nearest-event (right-hand panel) distribution functions. —— : data; – – – : envelope from 99 simulations of fitted model.

for this and for the data, visual inspection would suggest considerably fewer than the 26 or so clusters implied by the estimate $\hat\rho = 25.6$; the eye is a poor judge of the extent to which formally distinct clusters coalesce.

From a biological viewpoint, the fitted value of ρ implies that the mean number of mature redwoods in an area of around $500m^2$ is about 26. This seems an improbably large value, although Strauss (1975) gives no indication of the age at which the previous generation was felled. Assessment of the sampling distribution of $(\hat\rho, \hat\sigma)$ is therefore particularly relevant. Figure 7.5 shows the least squares estimates obtained from 100 simulated realizations of the fitted model. Note in particular the very wide range of values of $\hat\rho$ and the negative correlation between $\hat\rho$ and $\hat\sigma$.

Diggle (1978) fitted a different Poisson cluster process to these data, in which the radially symmetric Normal dispersion of offspring was replaced by a uniform distribution over a disc of radius σ centred on the corresponding parent. This gave an equally good fit, and comparable estimates $(\hat\rho, \hat\sigma)$. This example therefore illustrates that it can be all too easy to fit a model to sparse, strongly aggregated data. In the present context, a model incorporating inhibition between parents might be biologically more plausible, but in the absence of specific information on this point it would be difficult to justify the fitting of a more complex model to such a small set of data.

FIGURE 7.4
A realisation of the model fitted to the redwood seedling data.

7.4.2 Bramble canes

Hutchings (1979) describes an investigation of the spatial pattern of bramble canes in a 9 metre square plot "staked out within a dense thicket of bramble". Living canes were classified as newly emergent, one-year old or two-year old. Here, we consider only the newly emergent canes.

We first analyse the pattern of the 359 newly emergent canes, shown in Figure 7.6. For these data, Hutchings detects aggregation using the uncorrected form of the Clark-Evans test. Although this is strictly invalid, inspection of the data does strongly suggest an aggregated pattern and this is confirmed by analysis via the EDF's of nearest neighbour and point to nearest event distances.

For the remainder of the analysis we take 9 metres as the unit of measurement. Hutchings attributes the aggregated pattern to "vigorous vegetative reproduction." This suggests a Poisson cluster process as a provisional model. Using the same model and estimation procedure as for the analysis of the redwood seedling data reported in Section 6.3.1 we obtain parameter estimates $(\hat{\rho}, \hat{\sigma}) = (123.6, 0.012)$. The left-hand panel of Figure 7.7 shows the functions $\hat{K}(t) - \pi t^2$ for the data and from 99 simulations of the fitted model, together with the fitted function $K(t) - \pi t^2$. Even without formal testing, this clearly indicates a poor fit. The estimate for the data lies above all of those from the simulations for distances less than about 0.02, and below for distances between about 0.02 and 0.05. This suggests that the underlying process is generating pattern at two different scales, for which a plausible explanation might be a combination of the small-scale vegetative propagation and a larger-scale effect, perhaps environmental heterogeneity, whilst the Poisson cluster process can only capture a single scale of clustering, and the fit therefore tries to com-

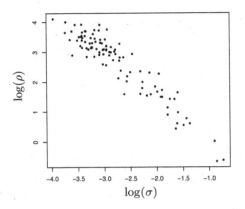

FIGURE 7.5
The empirical sampling distribution of $(\log \hat{\rho}, \log \hat{\sigma})$ for the model fitted to the redwood seedling data.

promise between the two. Note, incidentally, the very large sampling variation in $\hat{K}(t)$ at large distances t. Also, showing all of the simulated estimates as "shadows" gives a more complete picture of the sampling distribution of $\hat{K}(t)$ than plotting only the simulation envelopes.

In view of the above results, we now attempt to model the finer second-order structure apparent in the left-hand panel of Figure 7.7, by a combination of tight clustering at a small physical scale and a more diffuse form of aggregation. Specifically, we apply a thinning field $\{Z(x)\}$ to the Poisson cluster process described previously. For $\{Z(x)\}$ we use the example of Section 5.8.2, in which discs of radius δ are centred on the events of a Poisson process of intensity λ and $Z(x) = 1$ if x is covered by at least one such disc, $Z(x) = 0$ otherwise. According to (6.21) we need the mean and covariance function of $\{Z(x)\}$ in order to determine the second-order properties of this four-parameter model.

Note that $P\{Z(x) = 0\}$ is the probability that there are no disc centres within a distance δ of the point x and that $E[Z(x)] = P\{Z(x) = 1\} = 1 - P\{Z(x) = 0\}$. Since the number of disc centres in a region A follows a Poisson distribution with mean $\lambda|A|$, it follows that

$$E[Z(x)] = 1 - \exp(-\pi\lambda\delta^2). \tag{7.6}$$

Similarly, for two points x and y a distance u apart, $P\{Z(x) = Z(y) = 0\}$ is the probability that there are no disc centres within a distance δ of either x or y, and we deduce that

$$P\{Z(x) = Z(y) = 0\} = \begin{cases} \exp[-\lambda\{2\pi\delta^2 - A(u;\delta)\}] & : \quad 0 \le u < 2\delta \\ \exp(-2\pi\lambda\delta^2) & : \quad u \ge 2\delta \end{cases}$$

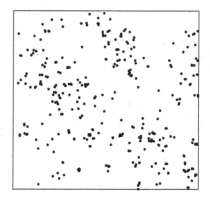

FIGURE 7.6
Locations of 359 newly emergent bramble canes in a 9 metre square plot
(Hutchings, 1978)

where $A(u, \delta)$ is the area of intersection of two discs, each of radius δ and whose centres are a distance u apart. This result allows us to evaluate the covariance function $\gamma(u)$ of $\{Z(x)\}$ because

$$\begin{aligned}
\text{Cov}\{Z(x), Z(y)\} &= \text{Cov}\{1 - Z(x), 1 - Z(y)\} \\
&= \text{P}\{Z(x) = Z(y) = 0\} - \text{P}\{Z(x) = 0\}\text{P}\{Z(y) = 0\}.
\end{aligned}$$

Routine manipulation gives the covariance function as

$$\gamma(u) = \begin{cases} \exp(-2\pi\lambda\delta^2)[\exp\{\lambda A(u; \delta)\} - 1] & : \quad 0 \leq u < 2\delta, \\ 0 & : \quad u \geq 2\delta. \end{cases} \tag{7.7}$$

Substitution of (7.6) and (7.7) into (6.21) gives $K(t)$ for our four-parameter model in a form which can easily be evaluated by numerical integration, namely

$$K(t) = K_0(t) + \mu^{-2} \int_0^t \gamma(u)K_0'(u)du,$$

where $\mu = 1 - \exp(-\pi\lambda\delta^2)$, $K_0(t) = \pi t^2 + \rho^{-1}[1 - \exp\{-u^2/(4\sigma^2)\}]$ and $K_0'(u) = 2\pi u + \exp\{-u^2/(4\sigma^2)\}/(2\rho\sigma^2)$.

Least squares estimates of the model parameters, again obtained using (7.1) with $t_0 = 0.25$, $c = 0.25$ and $w(t) = 1$ are $(\hat{\rho}, \hat{\sigma}, \hat{\lambda}, \hat{\delta}) = (587.6, 0.0036, 26.90, 0.1104)$. The right-hand panel of Figure 7.7 mirrors the left-hand panel, except that now the simulations are of the fitted four-parameter model. The fit is much improved, and warrants a formal assessment.

Figure 7.8 gives a graphical summary of the goodness-of-fit as assessed by

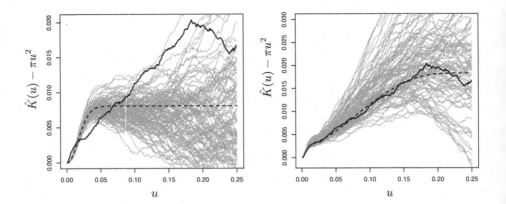

FIGURE 7.7

$K(t) - \pi t^2$ for newly emergent bramble canes (solid black line), fitted model (dashed black line) and 99 simulations of fitted model cluster process (thin grey lines). In the left-hand panel, the fitted model is a two-parameter Poisson cluster process; in the right-hand panel, it is four-parameter thinned Poisson cluster-process.

nearest neighbour and point-to-nearest-event distributions. The visual fit is tolerable, whilst the p-values for the associated tests based on (6.3) and (6.4), respectively, are 0.13 and 0.18.

Table 6.2 gives empirical standard errors of, and correlations amongst, the parameter estimates for the four-parameter model. These are calculated from the sample covariance matrix of parameter estimates obtained by applying the estimation procedure to 100 simulations of the fitted model. Note that none of the correlations are close to 1 in absolute value, which might suggest that the model is well-identified. However, the eigenvalues of the sample correlation matrix are 2.529, 1.105, 0.335 and 0.032. The first three principal components therefore account for approximately 96% of the variation in the joint sampling distribution of the four standardised parameter estimates, indicating that the distribution is effectively three-dimensional.

Figure 7.9 shows a realization of the fitted model, which can be compared with the data in Figure 7.6. A literal interpretation of the model would be that the Poisson cluster component describes the vegetative propagation of new shoots from old canes, whilst the thinning field distinguishes between parts of the plot which can or cannot sustain healthy growth. The truth is likely to be more subtle than this. Also, other models could doubtless be found that would fit the data equally well. However, the eigen-analysis reported above suggests that further elaboration of the model might be over-ambitious. Perhaps the only claim that should be made for this particular model is that it gives a

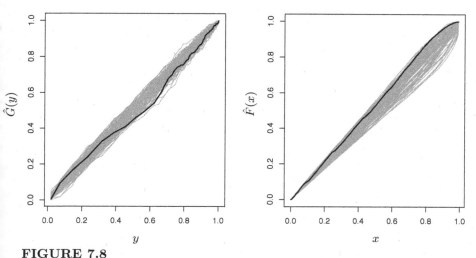

FIGURE 7.8

Goodness-of-fit for the four-parameter model fitted to the newly emergent bramble canes, using nearest neighbour (left-hand panel) and point-to-nearest-event (right-hand panel) distribution functions. The data are shown as thick black lines, simulations of the fitted model as thin grey lines.

tangible expression to the notion that the data incorporate two distinct scales of pattern.

We now extend the analysis to include the bivariate pattern formed by the 359 newly emergent and 385 one year old bramble canes. Because mortality is generally low in the first year of growth, these data can be interpreted as two successive cohorts of newly emergent canes. Figure 7.10 shows this bivariate data-set, and strongly suggests positive dependence between the two component patterns. Figure 7.11 shows that the three functions $\hat{K}_{ij}(t) - \pi t^2$ for the data are similar, at least for small t where sampling fluctuations in the $\hat{K}_{ij}(t)$ are also relatively small. In view of (4.13), this suggests random labelling as a possible model.

The four-parameter model which we fitted to the newly emergent canes is a Cox process, and within this parametric framework, random labelling of two component Cox processes corresponds to proportionality of the corresponding driving intensities, $\Lambda_2(x) \propto \Lambda_1(x)$, i.e. a *linked* Cox process as defined in Section 5.9.4. If the two components are indeed linked in this sense, then the parameters can be re-estimated from the superposition of the newly emergent and one year old canes. We obtain estimates $(\hat{\rho}, \hat{\sigma}, \hat{\lambda}, \hat{\delta}) = (745.2, 0.0034, 38.9, 0.0876)$. The magnitudes of the differences between these estimates and those obtained previously are compatible with the standard errors reported in Table 6.2.

To assess goodness of fit of the linked Cox process model, we first ex-

FIGURE 7.9
A realisation of the four-parameter model fitted to the newly emergent bramble canes

amine second-order properties. The left-hand panel of Figure 7.12 shows $\sqrt{\{\hat{K}_{12}(t)/\pi\}} - t$ for the bivariate data together with the corresponding functions calculated from 99 simulations of the fitted model. An informal visual assessment suggests a good fit. As in the univariate case, for a formal goodness-of-fit assessment it is preferable to use a test statistic that is not related to the method of parameter estimation. By analogy with the univariate case, a candidate for this task would be some kind of nearest neighbour analysis. For example, let $F_j(x)$ denote the distribution function of the distance from an arbitrary point to the nearest event of type j, and $F(x)$ the distance to the nearest event of either type. If the two component processes are independent,

TABLE 7.2
Empirical standard errors of, and correlations amongst, parameter estimates for model fitted to newly emergent bramble canes

Parameter	Point estimate	Standard error	Correlation with		
			$\hat{\sigma}$	$\hat{\lambda}$	$\hat{\delta}$
ρ	587.6	129.3	−0.58	0.38	−0.59
σ	0.0036	0.0004		−0.07	0.28
λ	26.9	14.33			−0.81
δ	0.1104	0.0444			

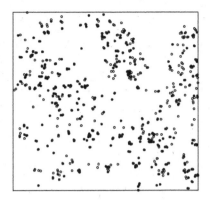

FIGURE 7.10
Locations of 359 newly emergent (solid dots) and 385 one-year-old (open circles) bramble canes in a 9 metre square plot (Hutchings, 1978).

then
$$F(x) = 1 - \{1 - F_1(x)\}\{1 - F_2(x)\},$$
More generally, we might consider the function
$$H_2(x) = \{1 - F(x)\} - \{1 - F_1(x)\}\{1 - F_2(x)\}$$
as measuring near-neighbour dependence between two components. A positive-valued function $H_2(x)$ indicates positive dependence in the sense that neighbouring type 1 and type 2 events are then, probabilistically speaking, closer together than in the case of independent components.

The right-hand panel of Figure 7.12 shows the empirical function $\hat{H}_2(x)$ for the bivariate data together with the envelope from 99 simulations of the fitted model. The data show stronger positive dependence near the origin than do the simulations of the model. A formal test using the statistic

$$\int_0^{0.2} \{\hat{H}_2(x) - \bar{H}(x)\}^2 dx \tag{7.8}$$

gives a *p*-value of 0.07.

We now add the two-year old canes to the analysis. A first candidate for a trivariate extension of the model is to a Cox process in which all three driving intensity processes are proportional, $\Lambda_3(x) \propto \Lambda_2(x) \propto \Lambda_1(x)$. To assess the fit of this trivariate linked Cox process, Figure 7.13 shows the trivariate version of $\hat{H}_2(x)$, namely

$$\hat{H}_3(x) = \{1 - \hat{F}(x)\} - \prod_{j=1}^{3} \{1 - \hat{F}_j(x)\},$$

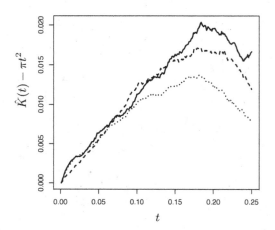

FIGURE 7.11

$\hat{K}_{ij}(t) - \pi t^2$ for newly emergent and one-year-old bramble canes: solid line corresponds to $K_{11}(t)$ (newly emergent), dashed line to $K_{22}(t)$ (one-year-old), dotted line to $K_{12}(t)$.

together with two sets of simulations. In the left-hand panel, the simulation model is the trivariate version of the linked Cox process, now with parameter estimates $(\hat{\rho}, \hat{\sigma}, \hat{\lambda}, \hat{\delta}) = 687.2, 0.0039, 37.8, 0.0890)$, whilst in the centre and right-hand panels the simulation model is random labelling, but without further parametric assumptions. In either case, a goodness-of-fit test using the trivariate analogue of (7.8) gives a p-value of 0.01, the most extreme result possible with 99 simulations. Visually, the lack of fit to the parametric Cox model is clear from the left-hand panel of Figure 7.13, whereas the centre panel suggests a superficially reasonable fit to the non-parametric random labelling hypothesis. The right-hand panel of Figure 7.13, which is simply a magnified version of the centre panel, shows a clear lack of fit, with the data-based function lying at or beyond the upper extreme of the simulation envelope throughout the reduced range of distances, $x \leq 0.025$.

Note, incidentally, how much tighter the simulation envelopes are in the non-parametric version of Figure 7.13. Conditioning on the data-locations in the superposition has greatly reduced the sampling variation in the test statistic. We conclude that the failure of the linked Cox process model to fit these data is not wholly attributable to our making inappropriate parametric assumptions; rather, the small-scale positive dependence amongst the three component pattern is stronger than can be explained by random labelling, and therefore by any trivariate Cox process.

There are only 79 two–year–old canes in the whole plot, because the majority of canes die after one year. Our non-parametric analysis has shown that

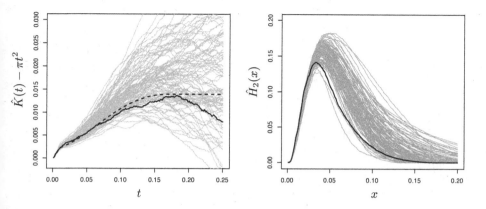

FIGURE 7.12
Left–hand panel: $\hat{K}_{12}(t) - \pi t^2$ for newly emergent and one-year-old bramble canes (thick black line) and for simulations of fitted linked Cox process model (thin grey lines). The fitted function $K_{12}(t) - \pi t^2$ is shown as a thick dashed line. Right–hand panel: $\hat{H}_2(x)$ for newly emergent and one-year-old bramble canes (solid black line) and for 99 simulations of fitted linked Cox process model (thin grey lines).

the data are inconsistent with any trivariate Cox process, and therefore that the observed pattern cannot be completely explained by micro-environmental variation. A more plausible explanation is that vegetative propagation in successive years generates a form of direct dependence over and above environmental heterogeneity.

7.5 Parameter estimation via goodness-of-fit testing

Another way of using summary descriptions in model-fitting is to exploit the duality between confidence intervals and significance tests. Suppose that we have a model indexed by a parameter vector θ, and a set of data presumed to be generated from the model but with the value of θ unknown. For any fixed value of θ we can test the goodness-of-fit of the model to the data. Then, the set of all values of θ which are not rejected by a test at the $100\alpha\%$ level constitutes a $100(1 - \alpha)\%$ confidence region for θ. If the individual tests are Monte Carlo tests, the method involves a direct search through a discretised version of the parameter space, and is only feasible for low-dimensional θ.

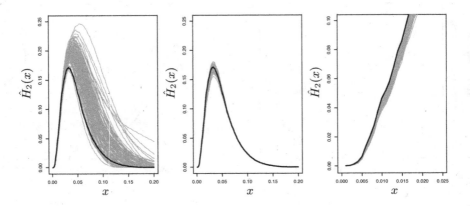

FIGURE 7.13

$\hat{H}_3(x)$ for newly emergent, one-year-old and two-year-old bramble canes (solid black line) and for 99 simulations (thin grey lines). In the left-hand panel, the simulation model is the fitted trivariate linked Cox process model, in the centre and right-hand panels the simulation model is random labelling.

7.5.1 Analysis of hamster tumour data

Figure 7.14, provided by Dr. W. A. Aherne (formerly of the Department of Pathology, University of Newcastle upon Tyne) shows the positions of the nuclei of 303 cells within an approximately 0.25mm square histological section of tissue from a laboratory-induced metastasising lymphoma in the kidney of a hamster. The diagram distinguishes two types of cells: 77 pyknotic nuclei, corresponding to dying cells, and 226 nuclei arrested in metaphase, corresponding to cells which have been "frozen" in the act of division. The background void is occupied by much larger numbers of unrecorded, interphase cells.

Figure 7.15 compares the second-order properties of the data with simulations of a pair of independent homogeneous Poisson processes as a provisional model. Again with the caveat about the small number of pyknotic cells, these again suggest a small-scale inhibitory effect, and are compatible with random labelling of the two cell types. We therefore fit an inhibitory pairwise interaction point process to the superposition of pyknotic and metaphase cells. Based on the discussion in Chapter 2, a natural goodness-of-fit statistic for an inhibitory process is

$$u = \int \{\hat{G}(y) - \bar{G}(y)\}^2 dy,$$

where $\hat{G}(\cdot)$ is the empirical distribution function of nearest neighbour distances for the data and $\bar{G}(\cdot)$ the corresponding average empirical distribution function from simulations of the simple inhibition process. Inspection of Figure 7.14

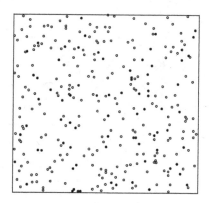

FIGURE 7.14
Locations of 303 cell nuclei in a hamster tumour; 77 pyknotic nuclei (solid dots); 226 metaphase nuclei (open circles)

shows several close pairs of cells. We therefore specified an interaction function with a non-strict form of inhibition, namely

$$h(u) = 1 - \exp\{-(u/\theta)^2\} : u \geq 0.$$

The left-hand panel of Figure 7.16 shows the resulting p-value as a function of θ, with the implied 95% confidence interval for θ highlighted. The right-hand panel of Figure 7.16 gives a more detailed summary of the goodness-of-fit when $\theta = 0.014 \approx 0.003mm$, corresponding to the mid-point of the confidence interval shown in the left-hand panel. The model fits well with, as already shown in the left-hand panel, a p-value of 0.72 for the goodness-of-fit test. However, other forms of interaction function might well give an equally good fit.

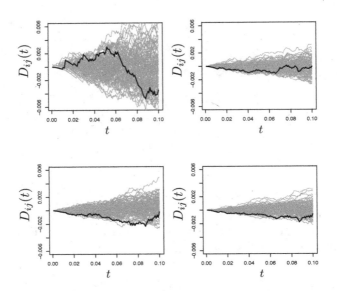

FIGURE 7.15
$D_{ij}(t) = \hat{K}_{ij}(t) - \pi t^2$ for the hamster tumour data, compared with 99 simulations of a pair of independent homogeneous Poisson processes. Upper-left panel shows $\hat{K}_{11}$ (pyknotic cells), upper-right panel $\hat{K}_{22}$ (metaphase cells), lower-left panel $\hat{K}_{12}$, lower-right panel $\hat{K}$ (superposition).

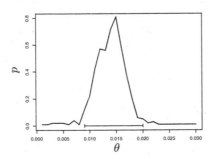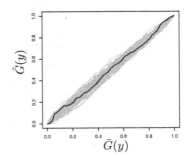

FIGURE 7.16
Left-hand panel: 95% confidence interval for interaction parameter θ in pairwise interaction process with $h(u) = 1 - \exp\{-(u/\theta)^2\}$, fitted to the hamster tumour data. Right-hand panel: goodness-of-fit for hamster tumour data, using the nearest neighbour distribution function. Thick black line is calculated from the data, thin grey lines from 99 simulations of pairwise interaction process with $\theta = 0.014$.

8

Model-fitting using likelihood-based methods

CONTENTS

8.1   Introduction

The likelihood function plays a fundamental role in both classical and Bayesian approaches to statistical inference. When more ad hoc methods, such as those described in Chapter 7, are used instead this is often for pragmatic reasons. The most obvious of these is that likelihood-based methods for most spatial point process models are notoriously intractable. However, this difficulty has to some extent been alleviated by recent developments in Monte Carlo methods of inference, including but not restricted to Markov chain Monte Carlo methods (Gilks, Richardson and Spiegelhalter, 1996).

8.2   Likelihood inference for inhomogeneous Poisson processes

An instance in which the likelihood function *is* tractable is the inhomogeneous Poisson process with intensity function $\lambda(x)$. Essentially, this is because the distribution associated with a partial realisation, $X = \{x_1, .., x_n\}$, of this process on a finite region A can be factorised as the product of a Poisson distribution with mean $\mu = \int_A \lambda(x) dx$ for the number of events n, and a set of mutually independent locations x_i whose common distribution has density $\lambda(x)/\mu$. Hence, the log-likelihood for $\lambda(\cdot)$ based on data X is

$$\{-\mu + n \log \mu - n!\} + \{\sum_{i=1^n} \log \lambda(x_i) - \log \mu\}$$

Simplifying, and ignoring the constant $n!$, this gives the log-likelihood as

$$L(\lambda) = \sum_{i=1}^{n} \log \lambda(x_i) - \int_A \lambda(x) dx. \tag{8.1}$$

In practice, this is most useful if $\lambda(x)$ can be specified through a regression model, for example a log-linear model

$$\log \lambda(x) = \sum_{j=1}^{p} \beta_j z_j(x) \tag{8.2}$$

where the $z_j(x)$ are spatially referenced explanatory variables. Cox (1972) calls this a *modulated* Poisson process.

The presence of the integral term on the right-hand side of (8.1) implies that, in order to fit the regression model (8.2), we need the explanatory variables $z_j(x)$ to be measured continuously throughout the study region, A. When the $z_j(x)$ are measured only at a finite number of points within A, we need to distinguish between observed and unobserved values of the underlying continuous surfaces $z_j(x)$, and the form of the likelihood function becomes more complicated. Briefly, consider a single explanatory variable, and partition the z-surface into observed and unobserved components, $z = \{z_o, z_u\}$ say. Then, the Poisson likelihood (8.1) would apply only if we observed the "complete" explanatory variable data z. To obtain a likelihood for the observed data, we need to specify a model for z, thereby defining the conditional distribution of z_u given z_o, and to eliminate z_u from the complete data likelihood (8.1) by integrating with respect to the conditional, $[z_u|z_o]$. How best to do this in practice is an open question. Rathbun (1996) develops an approximate approach, using geostatistical methods to interpolate from the observed z_o to the complete surface z. Note that modelling $z(x)$ stochastically turns the point process model into a Cox process. Stoyan and Ho (2008) consider the situation in which $z(x)$ is observed only at the points of the process, which they call an *intensity-marked Cox process*. Diggle, Menezes and Su (2010) use a similar model, but allow the possibility that values of $z(x)$ at the points of the process are measured with error. Standard geostatistical methods for making inferences about the process $Z(x)$, as described in Chilès and Delfiner (1999) or Diggle and Ribeiro (2007), assume that the spatially continuous process $Z(x)$ and the point process of locations at which $Z(x)$ is measured are stochastically independent. Diggle, Menezes and Su (2010) use their model to investigate how these standard methods perform when the data are generated by an intensity-marked process in which $Z(x)$ and the measurement locations are stochastically dependent.

For the Poisson case, i.e. when $z(x)$ is deterministic, Berman and Turner (1992) discuss different quadrature schemes for the integral term, and show how the model can then be fitted using standard generalized linear modelling software. Their method is implemented in the `Spatstat` package.

8.2.1 Fitting a trend surface to the Lansing Woods data

In Section 5.3 we used a nonparametric smoothing method to estimate intensity surfaces for the three major species groupings in the Lansing Woods data. Here, we use a parametric approach by specifying a trend surface for the logarithm of the intensity. A trend surface is a representation of a continuously varying surface by a polynomial in its two spatial coordinates. For example, a quadratic trend surface model is specified as

$$\log \lambda(x) = \alpha + \beta_1 x_1 + \beta_2 x_2 + \gamma_1 x_1^2 + \gamma_2 x_2^2 + \delta x_1 x_2, \qquad (8.3)$$

where x_1 and x_2 are the Cartesian coordinates of the location x. Table 8.1 summarises the results of fitting this model, and its linear and constant submodels, to each of the three major species groupings of the Lansing Woods data. In all three cases, formal likelihood ratio criteria would favour the quadratic over the linear or constant trend surface models. Notice, however, that the strength of the evidence in favour of the quadratic, as measured by the log-likelihood ratio, is greater for the hickories and maples than for the oaks, consistent with the relative amounts of spatial heterogeneity evident on visual inspection of the three data-sets (see Figure 2.11).

The results for the hickories and maples are shown graphically in Figure 8.1. In this example, the fitted trend surfaces capture the main features of the observed variation in spatial intensity reasonably well. Note in particular that the intensity of the hickories is relatively high where that of the maples is relatively low, and *vice versa*. However, as a general method, polynomial trend surface modelling is rather inflexible. In particular, when low-degree trend surfaces do not capture the essential features of the data, fitting higher-order surfaces tends to introduce artefacts, such as local modes where there is little or no data.

To assess the goodness-of-fit of the inhomogeneous Poisson process model,

TABLE 8.1

Maximised log-likelihoods for Poisson log-quadratic trend surface models fitted to the Lansing Woods data

Species	Maximised log-likelihood for trend surface of degree		
	0	1	2
hickories	3905.4	3943.3	3985.3
maples	2694.5	2747.6	2778.5
oaks	5419.9	5425.9	5440.4

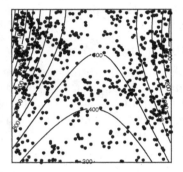 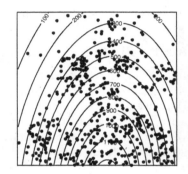

FIGURE 8.1

Fitted log-quadratic trend surfaces for the hickories (left-hand panel) and maples (right-hand panel) in Lansing Woods

we use the estimated inhomogeneous K-function, $\hat{K}_I(\cdot)$ defined by (4.26), plugging in the estimated log-quadratic surface $\hat{\lambda}(\cdot)$. The two panels of Figure 8.2 show $\hat{K}_I(t) - \pi t^2$ for the hickories and for the maples, in each case with simulation envelopes from 99 simulations of the fitted inhomogeneous Poisson process conditioned to produce a fixed number of events in the unit square. For both species, there is a clear lack of fit, which could be due either to non-Poisson behaviour (specifically, small-scale aggregation), or to inadequacy of the log-quadratic model.

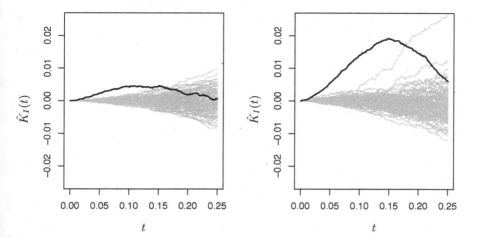

FIGURE 8.2

Estimates of $K_I(t)$ for the hickories (left-hand panel) and maples (right-hand panel) in Lansing Woods. Data are shown as thick black lines, simulations of the fitted log-quadratic inhomogeneous Poisson process as thin grey lines.

8.3 Likelihood inference for Markov point processes

As discussed in Section 5.7, Markov point processes are defined in terms of the joint density $f(X)$ for a configuration of events $X = \{x_1, ..., x_n\}$. This points to the likelihood function being an obvious tool for inference within this class. However, the normalising constant, represented by α in equation (6.14), is generally intractable.

An early response to the intractability problem was to replace the likelihood criterion by a *pseudo-likelihood* which, for a general multivariate distribution $f(x_1, ..., x_n)$ is defined as the product of the *full conditionals*, leading to the criterion

$$PL = \sum_{i=1}^{n} \log f(x_i|x_j, j \neq i).$$

Besag (1975) introduced the pseudo-likelihood as a method of parameter estimation for lattice processes. Besag, Milne and Zachary (1982) derived a point process version by showing that a pairwise interaction process could be defined as the limiting form of a binary lattice process when the lattice spacing shrinks to zero.

Subsequently, Monte Carlo methods for approximating the normalising constant were developed, although their implementation is not entirely straightforward and can be computationally burdensome. Partly for this reason, interest in the computationally less demanding pseudo-likelihood method was revived by Baddelely and Turner (2000). In the remainder of this chapter, we first consider the method of maximum pseudo-likelihood in more detail, before discussing Monte Carlo implementations of likelihood-based methods.

One issue that immediately arises in considering methods of inference for Markov point processes is whether or not to condition on the observed number of events, n. As discussed in Section 5.7, the distinction is not always innocuous, but is usually so for models that generate regular patterns. In developing specific estimation criteria we shall therefore take a pragmatic stance in choosing whether to treat n as fixed or random.

8.3.1   Maximum pseudo-likelihood estimation

Consider any process defined by the joint density $f(X)$ for a configuration consisting of a variable number of points, $X = (x_1, ..., x_n)$, in a spatial region A. We consider both n and the x_i to be random variables. The *conditional intensity* of a point at an arbitrary location u, given the realisation X of the process on $A - \{u\}$, is

$$\lambda(x; X) = \begin{cases} f(X \cup \{u\})/f(X) & : \quad u \notin X \\ f(X)/f(X - \{u\}) & : \quad u \in X \end{cases} \tag{8.4}$$

and the log-pseudo-likelihood function (Besag, 1978) is

$$PL_\lambda = \sum \log \lambda(x_i; X) - \int_A \lambda(u; X) du \tag{8.5}$$

For an inhomogeneous Poisson process, the conditional intensity reduces to the intensity, and the log-pseudo-likelihood (8.5) reduces to the log-likelihood (8.1).

More generally, for the class of pairwise interaction processes defined by (6.16),

$$\lambda(x_i; X) = \beta \prod_{j \neq i} h(||x_i - x_j||, \phi)$$

and, for $u \notin X$,

$$\lambda(u; X) = \beta \prod_{i=1}^{n} h(||u - x_i||), \phi).$$

Hence, writing $p(\cdot) = \log h(\cdot)$ for convenience, the log-pseudo-likelihood function for the model parameters (β, ϕ) is

$$PL(\beta, \phi) = n \log \beta + \sum_{i=1}^{n} \sum_{j \neq i} p(||x_i - x_j||; \phi) - \beta \int_A \exp\{\sum_{i=1}^{n} p(||u - x_i||, \phi)\} du.$$
$$\tag{8.6}$$

For any fixed value of ϕ, the value of β that maximises the pseudo-likelihood is given explicitly by

$$\tilde{\beta}(\phi) = n/ \int_A \exp\{\sum_{i=1}^{n} p(||u - x_i||, \phi)\}du. \tag{8.7}$$

Substitution of this expression back into the right-hand side of (8.6) gives the reduced log-pseudo-likelihood function,

$$PL_0(\phi) = n(\log n - 1) - n \log \int_A \exp\{\sum_{i=1}^{n} p(||u - x_i||, \phi)\}du + \sum_{i=1}^{n} \sum_{j\neq i} p(||x_i - x_j||; \phi). \tag{8.8}$$

This can be maximised numerically to give the maximum pseudo-likelihood estimator $\hat{\phi}$ and hence, by back-substitution into (8.7), the maximum pseudo-likelihood estimate $\hat{\beta} = \tilde{\beta}(\hat{\phi})$.

Note in particular that the log-pseudo-likelihood does not involve the awkward normalising constant α. Hence, maximum pseudo-likelihood estimation provides a computationally easy alternative to full maximum likelihood. Also, it can be implemented without the need for careful tuning of specialised Monte Carlo algorithms, as currently required for maximum likelihood estimation. In particular, Berman and Turner (1992) and Baddeley and Turner (2000) show how standard Poisson regression modelling software can be adapted to implement maximum pseudo-likelihood estimation for point process models including, but not restricted to, pairwise interaction point processes.

Diggle *et al.* (1994) suggested an edge-corrected version of the log-pseudo-likelihood function. The rationale for this is exactly parallel to that for the edge-correction described in Section 4.6 for estimating the K-function, namely that the summations of log-interaction terms on the right hand side of (8.8) ignore potential contributions from unobserved events outside the study region A. To address this, we replace (8.8) by

$$PL^*(\beta, \phi) = n(\log n - 1) - n \log \int_A \exp[\sum_{i=1}^{n} w(u, x_i)^{-1} p(||u - x_i||, \phi)du$$

$$+ \sum_{i=1}^{n} \sum_{j\neq i} w(x_i, x_j)^{-1} p(||x_i - x_j||; \phi), \tag{8.9}$$

where the weights $w(\cdot)$ are defined as in Section 4.6.1. A potential objection to (8.9) is that the edge-correction weights $w(x_i, x_j)^{-1}$ are unbounded. However, in practice it is only reasonable to attempt to estimate interactions that operate over a range of distances much smaller than the dimensions of the region on which the data are observed.

In practice, the integral in (8.8) or (8.9) must be evaluated numerically. In our applications, we use the simplest form of quadrature, replacing the integration by summation over a fine grid of equally spaced quadrature points to cover A.

8.3.2   Non-parametric estimation of a pairwise interaction function

In formulating pairwise interaction models for particular applications, it would be useful to have available a simple, non-parametric estimator for the interaction function $h(\cdot)$. This is needed because, in general, there is no simple algebraic relationship between $h(\cdot)$ and the already established summary descriptors based on second moment or nearest neighbour properties. Diggle, Gates and Stibbard (1987) propose a solution that combines Fourier methods and approximations from statistical physics, but their method is somewhat elaborate and difficult to automate. A simpler solution, suggested in Baddeley and Turner (2000), is to use maximum pseudo-likelihood in conjunction with a piece-wise constant specification for $h(\cdot)$, leading to a kind of indirect histogram estimator. Heikkinen and Penttinen (1999) also suggested estimating a piece-wise constant $h(\cdot)$, but using more computationally demanding, Monte Carlo likelihood-based methods of estimation.

8.3.3   Fitting a pairwise interaction point process to the displaced amacrine cells

In our earlier, non-parametric analysis of these data, reported in Section 4.7, we concluded that the *on* and *off* cells form independent processes with very similar second-order properties, including inhibitory behaviour at small distances. In view of this, our initial modelling strategy will be to formulate and fit a model to the *on* cells only, reserving the *off* cells for a goodness-of-fit assessment. Also, the inhibitory behaviour at small distances, together with the absence of any obvious longer-range heterogeneity or other form of aggregation, suggests that a pairwise interaction process is a reasonable candidate model.

For an initial, non-parametric estimate of the interaction function, we maximise the edge-corrected pseudo-likelihood criterion (8.9) in conjunction with a piece-wise constant interaction function. Figure 8.3 shows the resulting estimate of $h(\cdot)$. Its basic shape, increasing from 0 to approximately 1 over the range of distances 0 to 0.15, is characteristic of non-strict inhibitory interaction between the points.

For the particular estimate shown in Figure 8.3 the number and width of intervals on which the function is estimated were chosen after some preliminary experimentation. The quadrature points for evaluation of the integral term in the expression for the log-pseudo-likelihood were placed on an equally spaced 100 by 70 grid.

After inspection of this non-parametric estimate, we proceeded to fit a parametric model with interaction function

$$h(u) = \begin{cases} 0 & : \quad u < \delta \\ \{(u - \delta)/(\rho - \delta)\}^{\beta} & : \quad \delta \leq u \leq \rho \\ 1 & : \quad u > \rho \end{cases} \qquad (8.10)$$

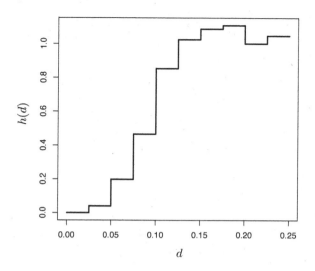

FIGURE 8.3
Non-parametric (step-function) estimate of the interaction function for the displaced amacrine *on* cells.

This particular parametric form has no special scientific status, but is simply a flexible family that provides a reasonable empirical fit to the form of interaction indicated by the preliminary, non-parametric estimate. Diggle and Gratton (1984) fitted the same model, but using curve-fitting methods of the kind described in Chapter 6. Here, we use maximum pseudo-likelihood to estimate the parameters δ, ρ and β. This involves a grid search in (δ, ρ)-space, with β optimised automatically at each value of (δ, ρ). Note that the minimum distance between any two of the *on* cells is approximately 0.032, hence the maximum pseudo-likelihood estimate of δ cannot be greater than 0.032. Again using a 100 by 70 grid of quadrature points for each evaluation of the integral term in the log-pseudo-likelihood we obtained estimates $\hat{\delta} = 0.016$, $\hat{\rho} = 0.121$ and $\hat{\beta} = 2.091$. Figure 8.4 compares this parametric estimate with the non-parametric estimate shown in Figure 8.3.

Figure 8.5 shows a profile of the log-pseudo-likelihood surface, maximised with respect to β for each of the plotted combinations of δ and ρ. The profile is a unimodal surface in the neighbourhood of the maximum pseudo-likelihood estimates, albeit somewhat flat in the δ-direction.

Figure 8.6 assesses the goodness-of-fit of the parametric model to the *off* cells, using nearest neighbour and point-to-nearest-event distribution functions. In both cases, there is a discrepancy between data and model as judged by the nearest neighbour distributions; the model generates patterns that are

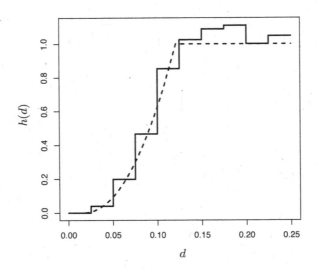

FIGURE 8.4
Parametric (dashed line) and non-parametric (step-function) estimates of the
interaction function for the displaced amacrine *on* cells.

more regular than the data. The corresponding Monte Carlo tests using the
statistics (7.4) and (7.3) give $p = 0.01$ and $p = 0.05$, respectively.

8.3.4   Monte Carlo maximum likelihood estimation

Recent developments in likelihood-based inference for point processes focus
on the use of Monte Carlo methods to circumvent problems of analytical intractability. In this section we describe an elegant method for Markov point
processes which was first used by Penttinen (1984), and subsequently developed in a more general context by Geyer and Thompson (1992). Geyer (1999)
describes how the method can be applied to spatial point processes including,
but not restricted to, Markov point processes.

For definiteness, we consider the fixed-n version of a pairwise interaction
point process with interaction function $h(d, \theta)$, where d denotes distance and
θ is the (possibly vector-valued) parameter to be estimated. We use $X = \{x_1, ..., x_n\}$ to denote a configuration of n events within some designated region
A. Then, the likelihood function for θ is

$$\ell(\theta) = f(X, \theta)/a(\theta) \tag{8.11}$$

where

$$f(X) = \prod_i \prod_{j \neq i} h(||x_i - x_j||, \theta).$$

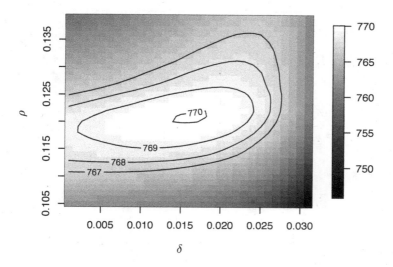

FIGURE 8.5
Profile log-pseudo-likelihood surface for the pairwise interaction point process fitted to the *on* cells. The plotted value at each point is the log-pseudo-likelihood maximised with respect to β, holding δ and ρ fixed.

The obstacle to direct evaluation of the likelihood is the normalising constant, $a(\theta)$. However, if we write $L(\theta) = \log \ell(\theta)$ and consider any fixed value θ_0 in the parameter space, it follows from (8.11) that

$$L(\theta) - L(\theta_0) = \log\{f(X,\theta)/f(X,\theta_0)\} - \log\{a(\theta)/a(\theta_0)\}. \qquad (8.12)$$

Observe that

$$
\begin{aligned}
a(\theta) &= \int f(X,\theta)dX \\
&= \int f(X,\theta) \times \frac{a(\theta_0)}{a(\theta_0)} \times \frac{f(X,\theta_0)}{f(X,\theta_0)} dX. \qquad (8.13)
\end{aligned}
$$

Now, define $r(X,\theta,\theta_0) = f(X,\theta)/f(X,\theta_0)$ and re-arrange the right-hand side of (8.13) to give

$$
\begin{aligned}
a(\theta) &= a(\theta_0) \int r(X,\theta,\theta_0)a(\theta_0)f(X,\theta_0)dX \\
&= a(\theta_0)\mathrm{E}_{\theta_0}[r(X,\theta,\theta_0)], \qquad (8.14)
\end{aligned}
$$

or equivalently

$$a(\theta/a(\theta_0) = \mathrm{E}_{\theta_0}[r(X,\theta,\theta_0)], \qquad (8.15)$$

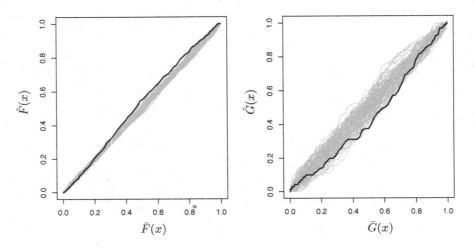

FIGURE 8.6
Goodness-of-fit assessment for displaced amacrine *off* cells, using point-to-nearest-event (left-hand panel) and nearest neighbour (right-hand panel) distribution functions. ——— : data; – – –: envelope from 99 simulations of fitted pairwise interaction point process.

where $\mathrm{E}_{\theta_0}[\cdot]$ denotes expectation with respect to the distribution of X when $\theta = \theta_0$. An immediate consequence of (8.14) is that if we can find a value θ_0 for which we can both evaluate $a(\theta_0)$ directly, and simulate realisations X, then in principle we can evaluate $a(\theta)$ approximately for any value of θ by using simulations, $X_1, ..., X_s$ say, under $\theta = \theta_0$ to obtain an empirical approximation to the expectation term.

More interestingly, substitution of (8.15) into (8.12) gives

$$L(\theta) - L(\theta_0) = \log\{f(X,\theta)/f(X,\theta_0)\} - \log\mathrm{E}_{\theta_0}[r(X,\theta,\theta_0)]. \qquad (8.16)$$

This suggests a family of algorithms for evaluating an approximate maximum likelihood estimator, $\hat{\theta}$. Since θ_0 is a constant, (8.16) implies that for any θ_0, the maximum likelihood estimator $\hat{\theta}$ maximises

$$L_{\theta_0}(\theta) = \log f(X,\theta) - \log\mathrm{E}_{\theta_0}[r(X,\theta,\theta_0)],$$

hence an approximation to $\hat{\theta}$ can be obtained by maximising

$$\hat{L}_{\theta_0,s}(\theta) = \log f(X,\theta) - \log s^{-1}\sum_{j=1}^{s} r(X_j,\theta,\theta_0) \qquad (8.17)$$

where $X_1, ... X_s$ are simulated realisations with $\theta = \theta_0$. Maximisation of (8.17) with respect to θ requires us only to be able to simulate the process at $\theta = \theta_0$.

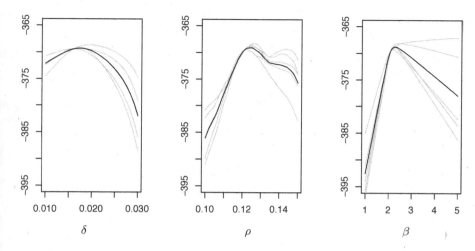

FIGURE 8.7
Monte Carlo log-likelihood functions for the displaced amacrine *on* cells, with interaction function (8.10), $\theta_0 = (\delta, \rho, \beta) = (0.016, 0.121, 2.091)$ and $s = 10$ simulations per log-likelihood evaluation. Thin grey lines show each of five independent replicates, thick black line the point-wise average of the five. Left-hand panel: $\hat{L}_{\theta_0,10}(\delta)$. Centre panel: $\hat{L}_{\theta_0,10}(\rho)$. Right-hand panel: $\hat{L}_{\theta_0,10}(\beta)$.

The Monte Carlo approximation to the expectation used in (8.17) may be numerically unstable with a poor choice of θ_0. However, the method of maximum pseudo-likelihood can be used to choose a value for θ_0, which we can consider as a first approximation to the maximum likelihood estimate, and this process can be iterated over successive approximations to $\hat{\theta}$. A precautionary comment is that even with modern computing facilities, the computational load of the Monte Carlo method can inhibit its routine use for fitting multi-parameter models to large data-sets.

8.3.5 The displaced amacrine cells re-visited

We now use the displaced amacrine cell data to illustrate the Monte Carlo maximum likelihood estimation procedure. To show how the method works, we first explore one-dimensional traces through the likelihood surface for the pairwise interaction model defined by (8.10), varying one of the three parameters whilst holding the other two fixed at their maximum pseudo-likelihood estimates.

Figure 8.7 shows the results for five replicate runs, in each of which the number of simulations used to evaluate (8.17) is $s = 10$ and $\theta_0 = (0.016, 0.121, 2.091)$. With such a small number of simulations, the compu-

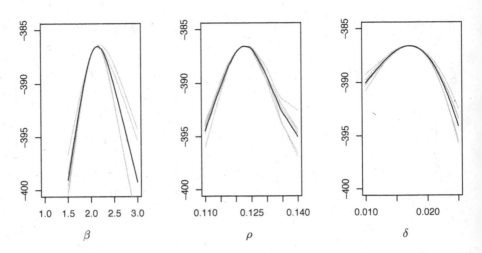

FIGURE 8.8
Monte Carlo log-likelihood functions for the displaced amacrine *on* cells, with interaction function (8.10), $\theta_0 = (\delta, \rho, \beta) = (0.017, 0.122, 2.100)$ and $s = 100$ simulations per log-likelihood evaluation. Thin grey lines show each of five independent replicates, thick black line the point-wise average of the five. Left-hand panel: $\hat{L}_{\theta_0,100}(\delta)$. Centre panel: $\hat{L}_{\theta_0,100}(\rho)$. Right-hand panel: $\hat{L}_{\theta_0,100}(\beta)$.

tation can be numerically unstable, as can be seen in the trace for δ and β in one of the five replicates.

Figure 8.8 gives the same information, but with $s = 100$ and θ_0 updated to $(0.017, 0.122, 2.100)$. The computations are now stable, and the Monte Carlo variation in the approximate log-likelihood is negligible in the vicinity of the maximum likelihood. This suggests that $s = 100$ gives an adequate approximation for point estimation. However, the divergence of the five replicates as θ moves away from the maximum likelihood estimate indicates that reliable interval estimation would need a larger value of s. Accordingly, Figure 8.9 shows the results obtained with $s = 1000$. The five replicates now show negligible divergence within the range relevant for calculating likelihood-based confidence intervals.

Taking the maximum likelihood estimates to be the values that maximise the average of the five replicates in each panel of Figure 8.9 gives maximum likelihood estimates $(\hat{\delta}, \hat{\rho}, \hat{\beta}) = (0.017, 0.122, 2.175)$. These differ only slightly from the maximum pseudo-likelihood estimates $(\hat{\delta}, \hat{\rho}, \hat{\beta}) = (0.016, 0.121, 2.091)$, and the fit to the nearest neighbour distribution is essentially as shown previously in Figure 8.6.

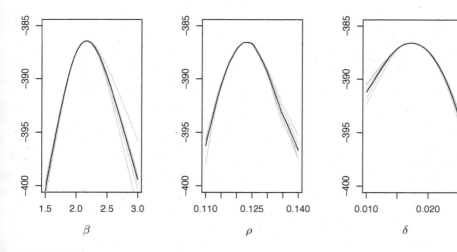

FIGURE 8.9

Monte Carlo log-likelihood functions for the displaced amacrine *on* cells, with interaction function (8.10), $\theta_0 = (\delta, \rho, \beta) = (0.017, 0.122, 2.100)$ and $s = 1000$ simulations per log-likelihood evaluation. Thin grey lines show each of five independent replicates, thick black line the point-wise average of the five. Left-hand panel: $\hat{L}_{\theta_0,1000}(\delta)$. Centre panel: $\hat{L}_{\theta_0,1000}(\rho)$. Right-hand panel: $\hat{L}_{\theta_0,1000}(\beta)$.

8.3.6 A bivariate model for the displaced amacrine cells

Until now, our working hypothesis for the amacrine cell data has been that the *on* and *off* cells are independent realisations of the same point process. We arrived at this working hypothesis by a non-parametric analysis as reported in Section 4.7, where we also concluded that the evidence pointed strongly towards the "separate layer" hypothesis for the underlying developmental biology. However, statistical independence cannot strictly be correct, as physical constraints prevent any two cells in the mature retina from occupying arbitrarily close positions in the retinal plane.

Figure 8.10 gives a schematic view of the transition from the immature to the mature retina. This prompted Eglen, Diggle and Troy (2005) to make a distinction between *statistical* and *functional* independence, the latter meaning that the only form of dependence between the two types of cell is a simple inhibitory constraint between cells of opposite type. Within the class of bivariate pairwise interaction point processes defined by equation (6.31) Section 5.9.5, the formal definition of functional independence is that the cross-interaction

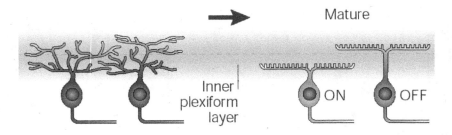

FIGURE 8.10
Schematic representation of the placement of displaced amacrine cells in the transition from the immature to the mature retina.

function, $h_{12}(u)$, takes the form

$$h_{12}(u) = \left\{ \begin{array}{lll} 0 & : & u \le \delta \\ 1 & : & u > \delta \end{array} \right. ,$$

with the special case $\delta = 0$ corresponding to functional and statistical independence.

Eglen, Diggle and Troy (2005) also used likelihood-based inference within the class of bivariate pairwise interaction point process models to test the assumption that the locations of the two types of cell are independent realisations of the same point process, i.e. that $h_{11}(u) = h_{22}(u)$. Their parametric model for the interaction functions was

$$h_{ij}(u) = \left\{ \begin{array}{lll} 0 & : & u \le \delta_{ij} \\ 1 - \exp[-\{u - \delta_{ij}\}/\phi_{ij}\}^{\beta_{ij}}] & : & u > \delta_{ij} \end{array} \right. .$$

This is different from the parametric form (8.10) but the difference is slight, and the two give very similar results. Eglen, Diggle and Troy (2005) also treated δ_{11} and δ_{22} as known constants, both equal to 0.015 based on the known physical size of a cell, but estimated δ_{12} because the vertical displacement between the two types of mature cells, as shown in the right-hand panel of Figure 8.10, would allow two cells of opposite type to be separated in the horizontal plane by less than their diameter. Using Monte Carlo likelihood ratio tests with $s = 1000$ replicates for each evaluation of the likelihood, they accepted the hypothesis $h_{11}(u) = h_{22}(u)$ (log-likelihood ratio 0.68 on two degrees of freedom, $p = 0.507$), rejected statistical independence against functional independence (log-likelihood ratio 2.65 on one degree of freedom, $p = 0.021$) and accepted functional independence against the general parametric form of $h_{12}(u)$ (log-likelihood ratio 0.15 on two degrees of freedom, $p = 0.861$). Maximum likelihood estimates for the fitted functional independence model, with $h_{11}(u) = h_{22}(u)$ as specified by equation (8.18) and $\delta_{11} = \delta_{22} = 0.015$ treated as known constants, are $(\hat{\phi}, \hat{\beta}, \hat{\delta}_{12}) = (0.085, 2.832, 0.008)$. Each evaluation of the Monte Carlo likelihood used $s = 1000$ replicate simulations with $\theta_0 = (\phi_0, \beta_0), \delta_{120} = (0.074, 2.92, 0.007)$. For numerical maximisation of the Monte Carlo log-likelihood we used the Nelder and Mead (1965) simplex algorithm as implemented in the R `optim()` function.

8.4   Likelihood inference for Cox processes

Recall from Chapter 6 that a Cox process is a Poisson process with stochastic intensity function $\Lambda(x)$. As a convenient shorthand, we write Λ for the complete set of values $\{\Lambda(x) : x \in A\}$. We assume that the model for $\Lambda(x)$ is indexed by a parameter θ. The likelihood function for θ associated with data $X = \{x_1, ..., x_n\}$ on a finite region A is therefore the expectation with respect to Λ of the Poisson process likelihood for a given Λ, which is $\exp(-\mu)\mu^{n-1} \prod \Lambda(x_i)/n!$ where $\mu = \int_A \Lambda(x)dx$. The overall intensity, μ, is usually of limited scientific interest and is easily estimated by the observed number of events per unit area. To estimate the remaining parameters of Λ, we can therefore condition on n, in which case the Poisson likelihood becomes

$$\ell(\theta, \Lambda) = c(\Lambda) \prod_{i=1}^{n} \Lambda(x_i), \tag{8.18}$$

where $c(\Lambda) = \{\int_A \Lambda(x)dx\}^{-n}$. The Cox process likelihood is then

$$\ell(\theta) = \mathrm{E}_\Lambda[\ell(\theta, \Lambda)]. \tag{8.19}$$

In the present context, the conditioning on n implies that (8.19) is unaffected if $\Lambda(x)$ is multiplied by an arbitrary constant; one implication of this is that in specifying a stationary parametric model for Λ, its expectation can be set at an arbitrary constant value, typically $\mathrm{E}[\Lambda(x)] = 1$. Also, the evaluation of (8.19) involves integration over the infinite-dimensional distribution of Λ. In principle, we can approximate Λ by a finite set of values $\Lambda(g_k) : k = 1, ..., N$, where the points $g_1, ..., g_N$ form a fine grid to cover A, but even so the high dimensionality of the implied integration appears to present a formidable obstacle to progress. One solution, easily stated but hard to implement robustly and efficiently, is to use Monte Carlo methods.

In essence, Monte Carlo evaluation of (8.19) consists of approximating the expectation by an empirical average over simulated realisations of some kind. A crude Monte Carlo method would use the approximation

$$\ell_{MC}(\theta) = s^{-1} \sum_{j=1}^{s} \ell(\theta, \lambda_j), \tag{8.20}$$

where $\lambda_j = \{\lambda_j(g_k) : k = 1, ..., N\}$ is a simulated realisation of Λ with parameter θ, on the set of grid-points g_k. In practice, this is hopelessly inefficient.

A more fruitful approach is to use an extension of the method described in Section 8.3.4, again building on results in Geyer (1999).

Imagine, temporarily, that we could observe both the point pattern, X, and the underlying intensity field Λ. The likelihood function would then take the form

$$\ell(\theta; X, \Lambda) = f(X, \Lambda; \theta)/a(\theta), \tag{8.21}$$

where $f(\cdot)$ is the un-normalised joint density of X and Λ, and

$$a(\theta) = \int \int f(X, \Lambda, \theta) d\Lambda dX \tag{8.22}$$

is the intractable normalising constant for $f(\cdot)$. The method described in Section 8.3.4 can be applied directly to give

$$a(\theta)/a(\theta_0) = \mathrm{E}_{\theta_0}[f(X, \Lambda; \theta)/f(X, \Lambda; \theta_0)], \tag{8.23}$$

where θ_0 is any convenient, fixed value of θ.

The function $f(X, \Lambda; \theta)$ in (8.21) is an un-normalised conditional density for Λ given X. Under this new interpretation, the corresponding normalised conditional density is $f(X, \Lambda; \theta)/a(\theta|X)$, where

$$a(\theta|X) = \int f(X, \Lambda, \theta) d\Lambda. \tag{8.24}$$

The argument leading to (8.23) can be repeated to give

$$a(\theta|X)/a(\theta_0|X) = \mathrm{E}_{\theta_0}[\{f(X, \Lambda; \theta)/f(X, \Lambda; \theta_0)\}|X], \tag{8.25}$$

We now acknowledge that only X is observed. The likelihood function is therefore obtained by integrating out the dependence on Λ on the right-hand side of (8.21), to give

$$\ell(\theta; X) = \left(\int f(X, \Lambda; \theta) d\Lambda \right) / a(\theta)$$

By substitution from (8.24), this can also be written as

$$\ell(\theta; X) = a(\theta|X)/a(\theta).$$

By the same token, for any fixed value θ_0,

$$\ell(\theta_0; X) = a(\theta_0|X)/a(\theta_0).$$

It follows that for any fixed value θ_0, the likelihood ratio between θ and θ_0 is

$$
\begin{aligned}
\ell(\theta; X)/\ell(\theta_0; X)\} &= \{a(\theta|X)/a(\theta)\}/\{a(\theta_0|X)/a(\theta_0)\} \\
&= \{a(\theta|X)/a(\theta_0|X)\}/\{a(\theta)/a(\theta_0)\}
\end{aligned}
$$

and the corresponding log-likelihood ratio is

$$L(\theta) - L(\theta_0) = \log\{a(\theta|X)/a(\theta_0|X)\} - \log\{a(\theta)/a(\theta_0)\}. \qquad (8.26)$$

Now write $r(X, \Lambda, \theta, \theta_0) = f(X, \Lambda, \theta)/f(X, \Lambda\theta_0)$ and substitute (8.23) and (8.25) into (8.26). This gives the log-likelihood for θ associated with the observed data X as

$$L(\theta) - L(\theta_0) = \log \mathrm{E}_{\theta_0}[r(X, \Lambda, \theta, \theta_0)|X] - \log \mathrm{E}_{\theta_0}[r(X, \Lambda, \theta, \theta_0)]. \qquad (8.27)$$

For any fixed value of θ_0, a Monte Carlo approximation to the log-likelihood, ignoring the constant term $L(\theta_0)$ on the left-hand side of (8.27) is therefore given by

$$\hat{L}(\theta) = \log\left\{ s^{-1} \sum_{j=1}^{s} [r(X, \lambda_j, \theta, \theta_0) \right\} - \log\left\{ s^{-1} \sum_{j=1}^{s} [r(X_j, \lambda_j, \theta, \theta_0) \right\} \qquad (8.28)$$

Note that in the second term on the right-hand side of (8.28) the pairs (X_j, Λ_j) are simulated joint realisations of X and Λ at $\theta = \theta_0$, whilst in the first term X is held fixed at the observed data and the simulated realisations λ_j are conditional on X.

Exactly as in Section 8.3.4, the result (8.28) shows that a Monte Carlo approximation to the log-likelihood function, and therefore to the maximum likelihood estimate $\hat{\theta}$, can be found by simulating the process only at a single value, θ_0. As there, the quality of the approximation rests on identifying a value θ_0 that is itself close in value to $\hat{\theta}$. Here, an additional complication is the need to simulate conditional realisations of Λ given X. Simple, direct simulation methods are available for unconditional simulation of X and Λ jointly, but conditional simulation of Λ requires Markov chain Monte Carlo methods, for which careful tuning is generally required to ensure convergence within a feasible run-length. Taylor *et al.* (2013) describe a Metropolis-adjusted Langevin algorithm for conditional simulation of a log-Gaussian Cox process, and an implementation of this algorithm in an R package, `lgcp`.

8.4.1   Predictive inference in a log-Gaussian Cox process

The definition of the Cox process makes it suitable as a model for point patterns that are thought primarily to be the result of spatial variation in intensity, rather than of stochastic interactions amongst the events. Within the Cox class, the log-Gaussian Cox process is particularly suited to applications in which the focus of scientific interest is in the realisation of the underlying intensity surface, rather than its parameters. The reason is the model's empirical character; its parameters have no obvious mechanistic interpretation, but simply describe how variable is the intensity surface, and how the correlation between pairs of values varies according to the spatial separation between their corresponding locations.

Suppose that we wish to use an observed partial realisation $\mathcal{X} = \{x_i \in A : i = 1, ..., n\}$ to find out about some aspect of the realisation of the underlying intensity field $\Lambda(x)$. Quite generally, let $\mathcal{T}$ denote the quantity of interest, called the *target*. In different contexts, $\mathcal{T}$ might be any of the following: a real-valued quantity, for example the maximum value of $\Lambda(x)$ within A; multi-dimensional, for example a digitised image of the surface $\Lambda(x)$ throughout A; a random set, for example the sub-region of A within which the value of $\Lambda(x)$ exceeds a threshold value of practical significance. Then, the formal solution to the prediction problem is the conditional distribution of $\mathcal{T}$ given the data, $\mathcal{X}$. If a point predictor is wanted, we can use a summary property of the predictive distribution such as its mean or median; note, in this context, that the mean, $\hat{\mathcal{T}} = \mathrm{E}[\mathcal{T}|\mathcal{X}]$ minimises the mean square error, $\mathrm{E}[(\hat{\mathcal{T}} - \mathcal{T})^2]$. A richer summary is a sequence of quantiles of the predictive distribution. This is particularly useful when $\mathcal{T}$ is the complete surface, $\{\Lambda(x) : x \in A\}$, whose predictive distribution can be examined interactively through an animation of a sequence of cumulative probability maps, $\{q(x, c) = \mathrm{P}(\Lambda(x) \leq c : x \in A\}$, for increasing values of $c > 0$.

Let θ denote the parameters of the model. In *plug-in* prediction, we use the conditional distribution of $\mathcal{T}$ given $\mathcal{X}$ treating a point estimate $\hat{\theta}$ as if it were a fixed, known quantity. In Bayesian prediction, we assign a prior distribution to θ and use the conditional distribution of $\mathcal{T}$ given $\mathcal{X}$ averaged over the posterior for θ. Specifically, the Bayesian predictive distribution is

$$[\mathcal{T}|\mathcal{X}] = \int [\mathcal{T}|\mathcal{X}; \theta][\theta|\mathcal{X}]d\theta. \qquad (8.29)$$

The averaging over the posterior in (8.29) typically, but not necessarily, gives a more conservative prediction than the plug-in, $[\mathcal{T}|\mathcal{X}; \hat{\theta}]$, because the plug-in ignores the uncertainty in the estimate $\hat{\theta}$.

8.4.2   Non-parametric estimation of an intensity surface: hickories in Lansing Woods

In Section 5.3 we used a kernel smoothing method for non-parametric estimation of an intensity surface. In Section 8.2 we considered a parametric version of the problem, fitting a log-quadratic intensity surface for a inhomogeneous Poisson process model by maximum likelihood. Here, we re-cast the problem as one of prediction within an LGCP model, using the hickories in Lansing Woods as our example; the analysis is taken from Diggle, Moraga, Rowlongson and Taylor (2013).

For this analysis, we re-scaled the data to lie in a square of side-length 100 and fitted an LGCP with intensity $\Lambda(x) = \exp\{\beta + S(x)\}$, where $S(x)$ is a Gaussian process with mean $-0.5\sigma^2$, variance σ^2 and correlation function $\rho(u) = \exp(-u/\phi)$; recall that under this parameterisation, $\mathrm{E}[\exp\{S(x)\}] = 1$, hence the unconditional intensity of the Cox process is $\lambda = \exp(\beta)$. We obtain preliminary estimates of $\theta = (\sigma, \phi)$ by minimising a weighted least squares

criterion,

$$D(\theta) = \int_0^{0.25} (\hat{K}(t)^{0.25} - K(t;\theta)^{0.25})^2 dt$$

as described in Section 7.2.1. The resulting estimates are $\tilde{\theta} = (0.50, 12.7)$. We used these to inform a pragmatic prior specification, with $\beta \sim N(0, 20)$, $\log \sigma \sim N(\log(\tilde{\sigma}), 0.1)$ and $\log \phi \sim N(\log(\tilde{\phi}), 0.05)$.

For the MCMC sampling, we used a burn-in of 30,000 iterations followed by a further 200,000 iterations, of which we retained every 200th iteration so as to give a weakly dependent sample of size 1000. Figure 8.11 compares the prior and posterior distributions of the three model parameters showing, in particular, that the data give only very weak information about the correlation range parameter, ϕ.

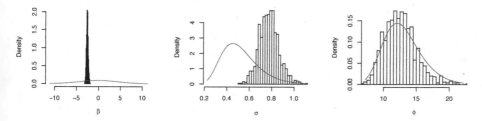

FIGURE 8.11
Prior (smooth curves) and posterior (histograms) distributions for the parameters β, σ and ϕ in the LGCP model for the hickory data.

The centre panel of Figure 8.12 shows the pointwise medians of the predictive distribution for the target, $\Lambda(x)$. This clearly identifies the pattern of the spatial variation in intensity, and is not dissimilar to the kernel-based estimate shown in Figure 5.3. The left-hand and right-hand panels show the pointwise predictive probabilities that the local intensity is less than half, and more than twice, the overall average, respectively. High probabilities in these plots indicate where there is strong evidence that the local intensity differs from the overall average by a factor of at least two in either direction.

The LGCP-based solution to the smoothing problem is arguably over-elaborate by comparison with simpler methods such as the kernel smoothing approach described in Section 5.3.1, especially if all that is required is an estimate of the intensity surface. Against this, arguments in its favour are that it provides a principled rather than an ad hoc solution, probabilistic prediction rather than point prediction, and an obvious extension to smoothing in the presence of explanatory variables by specifying $\Lambda(x) = \exp\{u(x)'\beta + S(x)\}$, where $u(x)$ is a vector of spatially referenced explanatory variables.

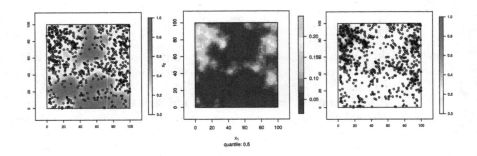

FIGURE 8.12
Predictive probabilities $P\{\exp[S(x)] < 1/2|\text{data}\}$ (left-hand panel), pointwise medians of the predictive distribution of $\exp\{\beta + S(x)\}$ (centre panel), predictive probabilities $P\{\exp[S(x)] > 2|\text{data}\}$ (right panel). Left and right panels also show data.

8.5   Additional reading

Likelihood-based Monte Carlo methods of inference for spatial point processes, including both classical and Bayesian approaches, are becoming more accessible through a combination of ever-increasing computational power and continuing theoretical developments. Early contributions include Ogata and Tanemura (1981, 1984, 1986) and Penttinen (1984). A good, if technically demanding, detailed treatment and literature review can be found in Møller and Waagepetersen (2002).

The author's view is that the more *ad hoc* methods of estimation will continue to be useful for some time because of their ease of implementation, and to provide good starting values for numerical optimisation of the likelihood. But for formal inference, they should eventually be replaced by likelihood-based methods as more efficient, reliable algorithms are developed and implemented in user-friendly software. Work by Havard Rue and colleagues on the use of Integrated Nested Laplace Approximations (Rue, Martino and Chopin, 2009; see also http://www.r-inla.org) reminds us that these algorithms may not need to use Monte Carlo methods.

9

Point process methods in spatial epidemiology

CONTENTS

9.1   Introduction

Epidemiology is concerned with the study of patterns of disease incidence and prevalence in natural populations, and the identification and estimation of risk factors associated with particular diseases. Traditionally, epidemiological studies only considered spatial risk factors at coarse geographical scales, for example comparing estimates of disease risk in different countries, or otherwise defined administrative regions. The advent of relatively precise post-code systems, together with the inclusion of post-coded information on place of birth, residence or death in disease registers and in census data, made it possible to consider much more detailed patterns of spatial variation in disease risk. For example, the UK post-code system is notionally accurate to an order of magnitude of tens of meters in urban areas, where each post-code typically identifies a single street. As a result, statistical methods have been developed to apply the ideas of spatial point processes to epidemiological data, specifically to the study of the observed pattern of disease in relation to possible environmental risk factors. In epidemiology, studies of this kind are often called *individual-level* studies. Studies that compare disease rates between different populations are usually called *area-level studies* or, somewhat quaintly, *ecological studies*.

Using point process methods to model the spatial pattern of disease is not an uncontroversial thing to do. At one level, it is obvious that allocating a person to a unique spatial location is no more than a convenient mathematical fiction. Even discounting long-term migration effects, most people move from place to place as they go about their daily business. Nevertheless, in the absence of direct, person-specific environmental dose monitoring, the location in which a person lives or works, according to context, may be the best available surrogate for the micro-environment to which they are principally exposed.

Another limitation of individual-level studies is that relevant collateral information, for example on demographic or socio-economic variables, is often only available on a larger spatial scale, for example at the level of counties or other administratively defined units. Against this, a powerful counter-argument in favour of individual-level studies is the well-known phenomenon of the *ecological fallacy* (Selvin, 1958), also called *ecological bias* (see, for example, Greenland and Morgenstern, 1989). This refers to the fact that effects of risk factors averaged over populations may differ, perhaps even qualitatively, from the corresponding effects at the individual level. Our aim in this chapter is to show how spatial point process methodology can be applied to several common problems in environmental epidemiology. We do not attempt to discuss the wider role which spatial statistical methods, including but not restricted to point process methods, can play in epidemiology. For overviews from this wider perspective see, for example, Elliott *et al.* (2000) or Waller and Gotway (2004).

The starting point for an individual-level analysis is a set of data giving the locations of all known *cases* of a particular disease within a designated study region A over a defined time-period. For example, the left-hand panel of Figure 9.1, based on data from Cuzick and Edwards (1990), shows the residential locations of 62 cases of childhood leukaemia diagnosed in the North Humberside region of the UK, in the years 1974 to 1982.

A feature of all data of this kind is that the spatial distribution of cases must to some extent reflect the spatial distribution of the underlying population. In the current example, the most obvious feature of the map is the concentration of cases in the city of Hull. Usually, patterns attributable solely to population distribution are not of interest, and it is therefore necessary to compare the case-map with a map of *controls* sampled from the underlying population at risk. The simplest case-control design is the completely randomised design, in which the controls are an independent random sample from the underlying population. The right-hand panel of Figure 9.1 shows a map of 143 controls sampled at random from the birth register for the North Humberside region over the years 1974 to 1982. Note the superficial similarity

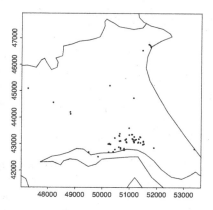 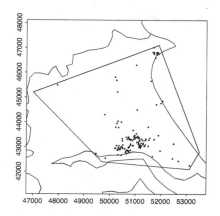

FIGURE 9.1

Residential locations at birth of 62 cases of childhood leukaemia in North Humberside, UK (left-hand panel), and of 143 controls sampled at random from the birth register (right-hand panel).

between the case and control maps, in the sense that both patterns broadly follow that of the underlying population at risk. The polygon superimposed on the control map is a crude approximation to the boundary of the North Humberside region, which we will use in the analysis reported in Section 9.2 below.

More sophisticated case-control designs involve stratification or matching; for example, a sample of controls may be constrained to show the same sex-ratio as the set of cases (*stratification* by sex), or controls may be paired with individual cases of the same age (*matching* by age). In what follows, we shall initially assume a completely randomised design. In Section 9.5 we discuss briefly how the associated statistical methods can be extended to cover stratified or matched designs, and point out why this may not be a good idea. For a general discussion of the arguments for and against matching in case-control study designs, see Breslow and Day (1980).

In the remainder of the chapter we shall discuss three classes of problem: investigation of spatial clustering of cases; non-parametric estimation of spatial variation in disease risk; and parametric modelling of elevation in risk near a point source of environmental pollution.

9.2 Spatial clustering

By *spatial clustering*, we mean a general tendency for cases to occur more closely together than would be compatible with random sampling from the population at risk. We emphasise that this is a description of the underlying disease process, rather than of the study region itself. The implication of clustering is that the conditional intensity of cases at an arbitrary location y, given a case at a nearby location x, is greater than the unconditional intensity of cases at y, i.e. clustering involves a form of dependence between cases.

Under the null hypothesis of no clustering, cases form a spatially random sample from the underlying population. By design, controls necessarily form a spatially random sample from this same population. Hence, no spatial clustering is equivalent to random labelling of the bivariate point process of cases and controls, and under this hypothesis the function

$$D(t) = K_{11}(t) - K_{22}(t) \tag{9.1}$$

is identically zero. More generally, $K_{22}(t)$ measures the degree of spatial aggregation of the population at risk, whereas $K_{11}(t)$ measures the cumulative effect of this same spatial aggregation together with any additional effect of clustering. Hence, $D(t)$ measures spatial clustering in the same way that $K(t) - \pi t^2$ measures the degree of spatial aggregation in a univariate process. We shall therefore develop a statistic to test the hypothesis of no clustering based on the corresponding empirical function,

$$\hat{D}(t) = \hat{K}_{11}(t) - \hat{K}_{22}(t),$$

where the case and control K-functions are estimated using (4.14).

In order to construct a formal test, we need to evaluate the null sampling distribution of $\hat{D}(t)$. In particular, although $D(t)$ itself is motivated by the theory of stationary spatial point processes, it would be inappropriate to assume stationarity in the present context because of the spatial heterogeneity inherent in human settlement patterns. We therefore turn to design-based inference, which uses the sampling distribution of $\hat{D}(t)$ induced by the random labelling process conditional on the observed superposition of cases and controls.

Diggle and Chetwynd (1991) use combinatorial arguments to show that under random labelling of cases and controls, $\mathrm{E}[\hat{D}(t)] = 0$ exactly. They also derive an explicit, albeit cumbersome, formula for the covariance $\mathrm{Cov}\{\hat{D}(t), \hat{D}(s)\}$. Based on these results, Diggle and Chetwynd (1991) suggest the test statistic

$$D = \int_0^{t_0} w(t)\hat{D}(t)dt \tag{9.2}$$

where $w(t) = \mathrm{Var}\{\hat{D}(t)\}^{-0.5}$. In applications, this requires a choice to be

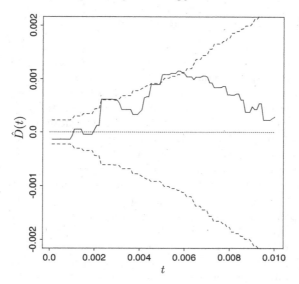

FIGURE 9.2
Second-order analysis of clustering for the North Humberside childhood leukaemia data. ——— : $\hat{D}(t)$ for observed data; − − − : plus and minus two standard errors under random labelling of cases and controls.

made for the upper limit of integration t_0. Our view is that this choice should be context-dependent. However, (9.2) implicitly down-weights large distances because the randomisation variance of $\hat{D}(t)$ increases with t. This makes the choice of t_0 less critical than it would otherwise be.

For an exact Monte Carlo test, we compare the observed value of D with values computed after independent random re-labellings of the cases and controls. A Gaussian approximation is also available if required, using the known form for the covariance structure of $\hat{D}(\cdot)$.

9.2.1 Analysis of the North Humberside childhood leukaemia data

Figure 9.2 shows the resulting analysis of the North Humberside childhood leukaemia data. The unit of distance is 100km. At distances of the order of several hundred meters the empirical function $\hat{D}(t)$ drifts close to or beyond the upper limit of two point-wise standard errors under the null hypothesis of no clustering. The p-value of an exact Monte Carlo test based on the statistic (9.2) is $p = 0.14$, whereas the Normal approximation gives $p = 0.11$. We conclude that there is only very slight evidence of spatial clustering in these data, and that any clustering which may occur operates on a spatial scale of the order of several hundred meters.

9.2.2 Other tests of spatial clustering

In this context, there is no compelling reason to use the Diggle-Chetwynd statistic D. There is an extensive literature on tests of spatial clustering for epidemiological data, somewhat reminiscent of the burgeoning of tests of complete spatial randomness for ecological data forty or more years ago, as reviewed in earlier chapters. The idea of using case-control data, and the randomisation distribution induced by the case-control design, to test for spatial clustering was introduced in Cuzick and Edwards (1990). Alexander and Boyle (1996) report on an empirical comparison amongst a number of different tests, based on their ability to detect clustering in a number of synthetic data-sets.

Methods for detecting specific clusters of disease are sometimes called *focused tests*, to distinguish them from tests of clustering as described above. A good example, with a carefully argued rationale for their use in practice, is Besag and Newell (1991). In effect, these methods operate by testing whether particular concentrations of cases are statistically significant, with appropriate modifications to allow for the implicit multiple testing. Other contributions to this area of research include Tango (1995, 2000), Anderson and Titterington (1997), Kulldorff (1997, 1999), Williams *et al.* (2001), Duczmal, Kulldorff and Huang (2006), Kulldorff *et al.* (2006) and Kulldorff *et al.* (2007).

Our preference for the Diggle-Chetwynd statistic stems from its roots in general summary statistics for point processes, rather than its performance in any specific power comparisons. In particular, the interpretation of $K(t)$ as a scaled expectation means that we can in turn interpret $D(t)$ as a scaled expected number of excess cases within distance t of a reference case, by comparison with a completely random pattern of disease incidence in the underlying population. Hence, at least in principle, it is possible to use $\hat{D}(t)$ not only to test the null hypothesis of no spatial clustering, but also to estimate the size and spatial scale of clustering if present. On the other hand, an implicit limitation of $\hat{D}(t)$ is that it is intended only to estimate general tendencies to clustering on spatial scales that are small relative to the dimensions of the study region. For a test of spatial clustering as here defined, the null hypothesis under test is that disease risk is spatially constant *and* cases occur independently, whereas the implicit alternative is that risk remains constant, but cases are dependent. When the suspected alternative is that risk varies spatially, either focused tests or regression methods may be more suitable. Focused tests can be viewed as methods for identifying unexplained peaks in the underlying risk surface, whereas regression methods are useful for investigating broad spatial trends in disease risk.

9.3   Spatial variation in risk

By *spatial variation in risk*, we mean that the case and control intensity functions are not proportional. Specifically, let $r(x)$ denote the probability that a person at location x will be a case. Then, adopting the usual convention that controls must be non-cases, the respective intensity functions of cases and controls are

$$\lambda_1(x) = r(x)\lambda(x) \qquad (9.3)$$

and

$$\lambda_2(x) = c\{1 - r(x)\}\lambda(x), \qquad (9.4)$$

where $\lambda(x)$ is the intensity of the underlying population and c is a constant determined by the study design. It follows that $\lambda_1(x)$ and $\lambda_2(x)$ are proportional if and only if $r(x)$ is constant. The function $r(x)$ is called the *risk surface*.

In contrast to spatial clustering, spatial variation in risk *is* a description of the study region, under the implicit assumption that cases of disease occur independently of one another. As we have seen in earlier chapters, it can be difficult or even impossible to sustain an empirical distinction between a process of dependent events in a homogeneous environment and one of independent events in a heterogeneous environment. To emphasise this in the present context, the null hypothesis of no spatial variation in risk, $r(x) = r$, is equivalent to random labelling of cases and controls, which is also the hypothesis of no spatial clustering. Thus, spatial clustering and spatial variation in risk represent different alternatives to the same null hypothesis.

Conditional on the intensity surface $\lambda_1(x)$, and under the assumption that cases occur independently, the case map is a realisation of an inhomogeneous Poisson process. Controls occur independently by design. It follows that conditional on the intensity surface $\lambda_2(x)$, the control map is a realisation of a second, independent Poisson process. Also, it follows from (9.3) and (9.4) that

$$\lambda_1(x)/\lambda_2(x) = c^{-1}r(x)/\{1 - r(x)\}. \qquad (9.5)$$

This shows that, up to a multiplicative constant, disease odds $r(x)/\{1-r(x)\}$, and hence the risk surface $r(x)$, can be estimated non-parametrically via non-parametric estimates of the two intensity functions $\lambda_j(x)$. Specifically, we can estimate the risk surface by substituting into (9.5) kernel estimates of the $\lambda_j(x)$ as discussed in Section 8.2.

In order to choose values of h for the kernel estimates of the $\lambda_j(x)$, we could use the method described in Section 8.2. However, in the non-parametric setting, there is no reason to suppose that optimal values of h for separate estimation of the two functions $\lambda_j(x)$ will be optimal for their ratio. Kelsall and Diggle (1995a) show that the asymptotically optimal estimator with respect to mean square error is achieved by using equal values of h in the numerator and denominator, irrespective of the numbers of cases and controls.

A second method of estimating the risk surface non-parametrically is motivated by the following observation. Consider two independent Poisson processes with respective intensities $\lambda_1(x)$ and $\lambda_2(x)$. Then, the superposition of the two is also a Poisson process, with intensity $\lambda_1(x) + \lambda_2(x)$. In this superposition, define a binary random variable Y_1 to take the value 1 or 0 according to whether the ith event in the superposition is an event of the first or the second component process. Then, conditional on the superposition the labels Y_i are mutually independent with

$$P(Y_i = 1) = \lambda_1(x_i)/\{\lambda_1(x_i) + \lambda_2(x_i)\}. \tag{9.6}$$

If we now substitute from (9.3) and (9.4) into the right hand side of (9.6), we obtain

$$\log\{P(Y_i = 1)/P(Y_i = 0)\} = -\log c + \log[r(x_i)/\{1 - r(x_i)\}]. \tag{9.7}$$

It follows from (9.7) that we can estimate the log-odds of disease, up to an additive constant, by using a non-parametric logistic regression model for the binary responses Y_i. This approach has the important advantage over the kernel density ratio estimator that it is easily extended to incorporate covariate information attached to individual cases and controls. Specifically, if we define the log-odds function $\ell(x) = \log[r(x)/\{1 - r(x)\}]$ and let z_i denote a covariate vector for the ith individual (case or control), then a semi-parametric model to identify residual spatial variation after adjusting for covariate effects is

$$\log\{P(Y_i = 1)/P(Y_i = 0)\} = \alpha + z_i'\beta + \ell(x_i). \tag{9.8}$$

Note in particular that the z_i could include spatial effects, such as a measure of social deprivation, or non-spatial effects such as the age or sex of the ith individual.

If the function $\ell(x)$ in (9.8) were to be specified parametrically, a natural approach to parameter estimation would be to maximise the log-likelihood,

$$L = \sum y_i \log p_i + (1 - y_i) \log(1 - p_i) \tag{9.9}$$

where $p_i = P(Y_i = 1)$ as specified by (9.8). In the non-parametric or semi-parametric setting an alternative method is to maximise a cross-validated version of the log-likelihood, as suggested in Diggle, Zheng and Durr (2005). Let $p(x)$ denote the estimated probability that an event at the location x is a case, using a non-parametric smoothing method with bandwidth h. Then, the cross-validated log-likelihood for h is

$$L_{CV}(h) = \sum y_i \log p^i(x_i) + (1 - y_i) \log(1 - p^i(x_i)) \tag{9.10}$$

where $p^i(x_i)$ denotes the estimated probability using all of the data except y_i.

The semi-parametric model (9.8) is an example of a generalised additive model (Hastie and Tibshirani, 1990; Wood, 2006). The R package `mgcv`, described in detail in Wood (2006), includes an implementation of spline-based methods for fitting models of this kind, embedded within a more general framework for semi-parametric generalised linear modelling.

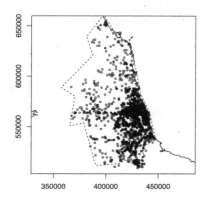 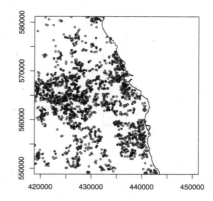

FIGURE 9.3
Residential locations (postcodes) of 761 cases of primary biliary cirrhosis in North East England (solid dots) and 3044 control locations selected by weighted random sampling from the postcode directory (open circles). The dashed line in the left-hand panel indicates the approximation to the boundary of the study-region used in the analysis of the data. The right-hand panel shows the densely populated part of the study-region in more detail.

9.3.1 Primary biliary cirrhosis in the North East of England

Prince et al. (2001) describe an analysis of data on liver cirrhosis in an area of the North East of England as shown in Figure 9.3. The 761 case-locations, shown as solid dots, are the residential locations of all incident and prevalent cases of definite or probable primary biliary cirrhosis alive between January 1987 and December 1994 amongst residents of a study-area defined by the boundaries of six health areas: Northumberland, Sunderland, North Durham, South Durham, South Tyneside, and North Tyneside. The 3044 control locations, shown as open circles, were selected randomly from a list of full UK postcodes within the study area. The probability for selecting each postcode was weighted by the number of drop-off points per postcode.

Using randomly selected postcodes as controls is problematic because it fails to distinguish between residential and commercial areas. Weighting by the number of drop-off points alleviates, but does not completely resolve, the problem because large commercial premises would typically have their own post-code with a single drop-off point, whereas in residential areas a single post-code would typically correspond to a complete street in urban areas, with one drop-off point per household. With this caveat, we now estimate a spatially varying risk surface, $r(x)$, for these data using the binary regression formulation (9.7). Figure 9.4 shows the cross-validated log-likelihood (9.10).

Using the bandwidth $h = 6.0$km that maximises the cross-validated likelihood, we obtain the estimated relative risk surface shown in Figure 9.5. A global test of the hypothesis of constant risk is overwhelmingly rejected

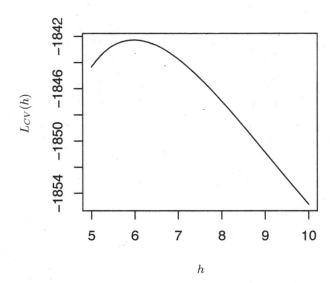

FIGURE 9.4
Cross-validated log-likelihood criterion for choice of band-width in non-parametric estimation of the relative risk surface for primary biliary cirrhosis in the North East of England.

($p = 0.001$ by a Monte Carlo test with 999 simulations). The contour line in Figure 9.5 identifies the sub-region, consisting of essentially the Newcastle-Gateshead conurbation, in which the local relative risk is significantly higher, at the conventional 5% level of significance, than the region-wide average. The obvious inference, that risk is higher in this sub-region than elsewhere in the region, needs to be tempered by the fact that the local tests are more powerful in sub-regions of relatively high population density. Note, however, that the highest point estimates of relative risk also fall in this same sub-region.

9.4   Point source models

The question of spatial variation in risk arises very directly when it is suspected that adverse effects on health are caused by a specific source of environmental pollution. A much-studied, and controversial, example in the UK has been the investigation of unusually high incidences of childhood cancers near nuclear installations. See, for example, Cook-Mozaffari et al. (1989) and Gardner (1989).

   In this more structured setting, it is reasonable to contemplate parametric

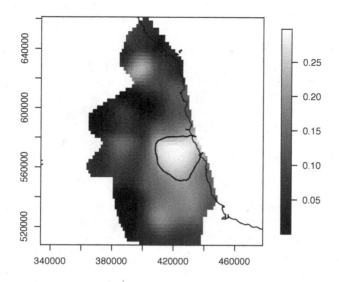

FIGURE 9.5
Estimated relative risk surface for primary biliary cirrhosis in the North East of England. The scale denotes the probability, at each location x in the study-region, that a case or control at location x is a case.

modelling of the risk surface in relation to the postulated source. This leads to a general model in which controls and cases form independent Poisson processes with respective intensities $\lambda_2(x)$ and $\lambda_1(x) = \phi\lambda_2(x)\rho(x)$ where $\lambda_2(x)$ is of unspecified form, $\rho(x)$ is given by a parametric model and ϕ is a nuisance parameter that relates to the relative numbers of cases and controls. Using the same argument as in Section 9.3, by considering case-control labels conditional on locations we can convert the Poisson process model to a binary regression model with spatially dependent probabilities

$$p(x) = \lambda_1(x)/\{\lambda_1(x) + \lambda_2(x)\} = \phi\rho(x)/\{1 + \phi\rho(x)\}, \qquad (9.11)$$

thereby eliminating the nuisance function $\lambda_2(x)$.

We first consider models that depend only on distance from a point source. Hence, the risk at a location x is proportional to some function $\rho(||x - x_0||)$ where x_0 is the location of the source and $|| \cdot ||$ denotes distance.

The simplest possible such model postulates an elevation in risk within some critical distance, δ, hence

$$\rho(u) = \begin{cases} 1+\alpha & : & u \leq \delta \\ 1 & : & u > \delta. \end{cases} \qquad (9.12)$$

In practice, this model is often used with a subjectively chosen value for δ

(Elliott et al., 1992). The resulting analysis is extremely simple. In this setting, case-control locations can be reduced to a 2×2 contingency table of events classified by their case-control designations and their distance from the source being greater or less than δ. The test for elevated risk within the selected distance threshold is then a standard comparison of two binomial proportions, and the corresponding empirical proportions of cases, p_1 and p_2 say, from events below and above the distance threshold, estimate $1/\{1 + \phi(1 + \alpha)\}$ and $\phi(1 + \alpha)/\{1 + \phi(1 + \alpha)\}$, respectively.

Lawson (1989), Diggle (1990) and Diggle and Rowlingson (1994) used an isotropic Gaussian model,

$$\rho(u) = 1 + \alpha \exp\{-(u/\delta)^2\}. \tag{9.13}$$

As in the previous model (9.12), the parameter α measures the elevation in risk at the source, whereas the second parameter δ now measures the rate at which risk decays smoothly with increasing distance towards a background level represented by $\rho(u) = 1$. Whilst there is no particular theoretical justification for assuming the Gaussian shape, a smoothly decaying risk function will be qualitatively sensible in many applications. Ideally, the form of model should be suggested by the practical context, for example to correspond to the behaviour of a plume of dispersing pollutant. This analogy immediately raises the possibility that the pattern of elevation in risk may have a directional component. As a simple example of how a directional effect might be incorporated, Figure 9.6 shows examples of a directional model with

$$f(d, \theta) = 1 + \alpha \exp(-[d \exp\{\kappa \cos(\theta - \phi)\}/\beta]^2). \tag{9.14}$$

In (9.14), α represents the elevation in risk at the source, β the rate of decay of risk with distance from the source, ϕ the principal direction of the plume and κ the extent of directional concentration of the plume, hence each of the parameters has a tangible interpretation. Lawson (1993) modelled directional effects by including terms for $\cos(\theta - \phi)$ and $u \cos(\theta - \phi)$ in a log-linear formulation for $\rho(u)$. Rodrigues, Diggle and Assuncao (2010) used a non-parametric specification for $\rho(u)$.

In practice, point source models are rarely derived from theoretical arguments. More often, they are used simply as parsimonious, descriptive models. As in the non-parametric case, it is important to include adjustments for known available risk factors in order to avoid the detection of spurious spatial effects. For example, pollution sources are often sited in areas with generally higher than average social deprivation, which is known to be a risk factor for many diseases.

All of the specific point source models described above correspond formally to non-linear binary regression models for the case-control labels, and can accordingly be fitted using likelihood-based methods. However, the properties of maximum likelihood estimators and likelihood ratio tests may show irregular behaviour. This is discussed, for example, in Diggle, Elliott, Morris and Shaddick (1997).

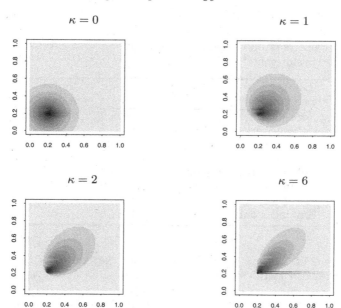

FIGURE 9.6
Examples of a directional model for elevation in risk around a point source.

9.4.1 Childhood asthma in north Derbyshire, England

Diggle and Rowlingson (1994) fit the isotropic Gaussian model to data from a case-control study of asthmatic symptoms in elementary schools in north Derbyshire, England. The study population consisted of children attending 10 schools in the area. Schools were stratified according to whether the head teacher had previously reported concern about the apparently high level of asthmatic symptoms in the school. Four potential sources were considered; here, we look only at two: a coking works, and the main road network. In the latter case, we used the distance between each child's residential location x and the nearest point on the road network as the distance measure in the model. Additional binary covariates for each child in the study indicated whether the household included at least one cigarette smoker, and whether the child suffered from hay fever. The overall risk was modelled multiplicatively, with separate terms for each of the two sources, and log-linear covariate adjustments for smoking, hay-fever and the prior stratification of the schools into two groups.

Likelihood ratio comparisons within this overall modelling framework are summarised in Table 9.1. Our conclusions are firstly that hay fever is biggest single risk factor and is overwhelmingly significant; secondly, that proximity to coke works shows a marginally significant increase in risk, with or without

TABLE 9.1
Deviances (twice maximised log-likelihoods) for various sub-models fitted to the north Derbyshire childhood asthma data

Risk factors included	Deviance	Number of parameters
None	-1165.9	2
Coking works	1160.7	4
Coking works, main roads	1160.6	6
Coking works, smoking	1159.4	5
Coking works, hay fever	1127.6	5
Hay fever only	1132.5	3

prior adjustment for hay fever. For example, the comparison between the last two lines of the table gives a chi-squared value of 4.9 on 2 degrees of freedom to test the association with the coke works after adjusting for hay fever ($p = 0.087$). There is no evidence of significant association with main roads or with cigarette smoking.

9.4.2   Cancers in North Liverpool

We now present the results of an investigation that used both parametric and non-parametric approaches to the estimation of spatial variation in risk. The investigation concerned the spatial distribution of cancer cases in an area of North Liverpool, UK, in which specific concerns had been expressed about a possible elevation in risk near the site of a now-disused hospital incinerator. The results presented here are extracted from Ardern (2001).

The study used an unmatched case-control design. The case locations consisted of the residential post-codes of all known cases of cancer diagnosed between 1974 and 1988. Adult cancers were classified into seven types, as listed in Table 9.2.

A random sample of 10,000 control locations was drawn from a database of general practitioner registrations within the study area, as shown in Figure 9.7. Also shown in Figure 9.7 is the location of the former incinerator. Covariate information attached to each case and to each control included the individual's age and sex, and the Townsend index of social deprivation (Townsend *et al.*, 1988) for the census enumeration district that contained the individual's residential location.

The initial analysis consisted of fitting the isotropic model (9.13), including log-linear adjustments for age, sex and the Townsend index. For every type of cancer, after adjustment for covariate effects the association with distance from the incinerator was non-significant. Table 9.3 shows the esti-

TABLE 9.2

Numbers of North Liverpool cancer cases available for analysis

Cancer type	Number
Colorectal	1162
Lung	2345
Liver	70
Larynx and nasopharynx	126
Leukaemia and lymphoma	365
Soft tissue sarcoma	45
Other cancers	5828
Total	9941

TABLE 9.3

Parameter estimates of regression effects for age, sex and Townsend deprivation index (TI) in the North Liverpool cancer study, and p-values for associated likelihood ratio tests of significance.

Cancer type	Parameter estimate			p-value		
	age	sex	TI	age	sex	TI
Colorectal	0.076	−0.31	0.033	< 0.001	< 0.001	< 0.001
Lung	0.077	−0.97	0.086	< 0.001	< 0.001	< 0.001
Liver	0.062	−0.56	0.071	< 0.001	0.02	0.06
Larynx/nasopharynx	0.059	−1.34	0.125	< 0.001	< 0.001	< 0.001
Leukaemia/lymphoma	0.048	−0.30	0.043	< 0.001	0.005	0.01
Soft tissue sarcoma	0.045	−0.32	−0.048	< 0.001	0.29	0.28
All adult cancers	0.075	−0.14	0.036	< 0.001	< 0.001	< 0.001

mated covariate adjustment parameters and their statistical significance. As expected, the effect of age is highly significant and positive for all adult cancer types considered. The effect of sex is highly significant for colorectal, lung and larynx/nasopharynx cancers, with men at higher risk than women. Sex is less significant for liver and leukaemias/lymphomas and non-significant for soft tissue sarcomas, although this may be a reflection of the smaller sample sizes available for the less common cancer types. Social deprivation as measured by the Townsend index is highly significant for colorectal, lung and larynx/nasopharynx cancers, and less significant or non-significant for the remaining, less common types. This may again be a reflection of the smaller sample sizes available.

The conclusion from the initial analysis is therefore that there is no significant evidence of association between cancer risk and proximity to the lo-

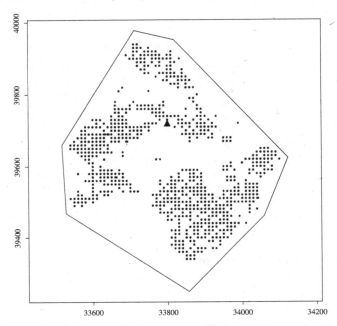

FIGURE 9.7
Study region, control locations (solid dots) and incinerator location (solid triangle) for the North Liverpool cancer study.

cation of the former incinerator. To investigate the possibility that there may be unexplained spatial variation in risk that is not associated with the incinerator, we used the semi-parametric model (9.8), again adjusting for age, sex and social deprivation as measured by the Townsend index. For the kernel smoothing term, we used a quartic kernel with a subjectively chosen bandwidth $h = 0.5$km. We applied this semi-parametric model only to the more common cancer types, as the kernel smoothing method is only effective with large sample sizes. Figures 9.8 to 9.10 show the resulting estimates of residual spatial variation in risk. In each case, the scale is logarithmic to base 2, hence each unit increase in the grey-scale corresponds to a doubling of estimated risk. The solid and dashed contours identify regions within which the local risk is pointwise significantly higher or lower, respectively, than the average for the whole study-area, at the conventional 5% level. All three cancer types show an area of apparently elevated risk close to the north-eastern boundary of the study-area. The p-values for an overall test of departure from constant residual risk are 0.05, 0.01 and 0.67 for colorectal, lung and leukaemia/lymphoma, respectively. This suggests that, at least for colorectal and lung cancers, the statistical significance of the elevated risk close to the north-eastern boundary is not merely a by-product of multiple testing.

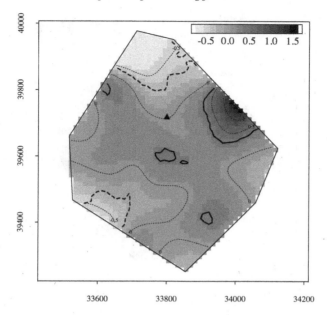

FIGURE 9.8
Estimated residual spatial variation in risk for colorectal cancers in the North Liverpool cancer study. The risk scale is logarithmic to base 2. Solid and dashed contours identify regions within which risk is point-wise 5% significantly higher or lower, respectively, than the average for the whole study-area.

9.5   Stratification and matching

9.5.1   Stratified case-control designs

All of the methods described above are easily adapted to stratified case-control studies. Provided the number of events within each stratum is sufficiently large, the analysis can be carried out separately within each stratum and the results pooled as and when appropriate. The precise form of pooling will depend on what supplementary assumptions are considered to be reasonable. We illustrate this for the specific case in which there are two strata, for example one for each sex.

We first consider how to modify the Diggle-Chetwynd test for spatial clustering when cases and controls can each be divided into two strata. Compute test statistics D_1 and D_2 within each stratum, as described in Section 9.2. If v_j denotes the null variance of D_j according to the Diggle-Chetwynd formula, then a suitable combined test statistic is given by $D = v_1^{-0.5} D_1 + v_2^{-0.5} D_2$. Under the reasonable assumption that the labelling processes operating in

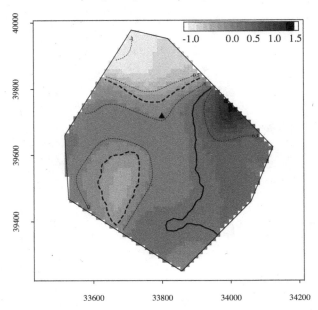

FIGURE 9.9
Estimated residual spatial variation in risk for lung cancers in the North Liverpool cancer study. The risk scale is logarithmic to base 2. Solid and dashed contours identify regions within which risk is point-wise 5% significantly higher or lower, respectively, than the average for the whole study-area.

the two strata are independent, the null expectation and variance of D are zero and 2, respectively. An approximate test follows by assuming a Normal sampling distribution for D, whilst an exact Monte Carlo test is available by jointly re-labelling cases and controls randomly within each stratum.

We now consider non-parametric estimation of a spatially varying risk surface when there are two strata. In this case, we use the generalized additive model formulation (9.8). The simplest way to incorporate strata into the analysis is then as a two-level factor to be added to the model as a main effect. If necessary, the stratum factor could then be allowed to interact with other terms in the model.

For the parametric modelling of elevated risk near a point source, the same basic strategy applies. We introduce the stratum label as a two-level factor, to be added to the regression model either as a main effect or as an interaction with other terms.

These methods of dealing with data in two strata extend in the obvious way to $k > 2$ strata. However, in practice this is only a useful strategy if the number of strata is small and the number of events within each stratum is large.

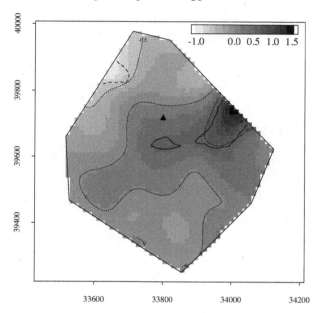

FIGURE 9.10
Estimated residual spatial variation in risk for leukaemias/lymphomas in the
North Liverpool cancer study. The risk scale is logarithmic to base 2. Solid
and dashed contours identify regions within which risk is point-wise 5% significantly higher or lower, respectively, than the average for the whole study-area.

9.5.2 Individually matched case-control designs

When strata are small, a different approach is needed. We consider here the
setting of individually matched data, whereby each case is associated with a
set of k controls matched to the corresponding case by the values of one or
more identifiable factors.

To investigate spatial clustering in this setting, Chetwynd *et al.* (2001)
evaluate the null expectation and covariance structure of $\hat{D}(t) = \hat{K}_{11}(t) - \hat{K}_{22}(t)$ for an individually matched case-control design. They show that when
$k = 1$, the null expectation of $D(t)$ is still zero, whereas when $k > 1$ it is
non-zero, perhaps substantially so. This suggests that we should modify the
the test statistic (9.2), for example to

$$D' = \int_0^{t_0} v(t)^{-0.5} \{\hat{D}(t) - \mu(t)\} dt$$

where $\mu(t)$ denotes the null expectation of $D(t)$, and $v(t)$ now denotes the
variance calculated from the randomisation distribution appropriate to the
individually matched design.

An intuitive explanation for the non-zero null expectation in the case $k >$

1 is that the matching variables may themselves be spatially non-neutral. For example, suppose that marginally the population, and hence the cases under the null hypothesis of no spatial clustering, are distributed completely at random, but that matched controls are likely to be spatially close to the corresponding case. Then, the cases would generate an estimate $\hat{K}_{11}(t) \approx \pi t^2$, but when $k > 1$ the controls would tend to fall in clumps of size k, leading to $\hat{K}_{22}(t) > \pi t^2$. This kind of effect can easily arise in practice because administrative, demographic or socio-economic factors, which are often used as matching variables, tend to be spatially non-neutral. A corollary is that individual matching can easily induce spurious spatial effects if the analysis fails to make allowance for the matching in the study design.

For non-parametric or parametric modelling of spatial variation in risk using the binary regression formulation, individual matching requires the basic form of the likelihood function introduced in Section 9.3 to be modified. Let $p(x)$ denote the probability that a person at location x is a case, and x_{ij} the location of the jth member of the ith matched case-control set with $i = 1$ identifying the case and $i = 2, ..., (k + 1)$ its matched controls. Then, the matched design is equivalent to a constraint that amongst the $k + 1$ members of any matched set, there is exactly one case. The probability that the case is the member at location x_{i1} is therefore given by

$$p_j = p(x_{i1}) / \sum_{i=1}^{k+1} p(x_{ij}).$$

The corresponding log-likelihood for n matched sets is given by

$$L^* = \sum_{j=1}^{n} \log p_j. \tag{9.15}$$

The corresponding expression for a randomised case-control design is given by (9.9).

Diggle, Morris and Wakefield (2000) discuss inference based on the log-likelihood (9.15) in the specific context of point source models. Jarner, Diggle and Chetwynd (2002) discuss non-parametric estimation of the risk surface. As in the case of spatial clustering, when matching variables are spatially non-neutral they introduce an ambiguity into precisely what is being estimated as a spatial effect. For this reason, when spatial variation is of particular interest, we would recommend dealing with measured risk factors by regression adjustments where possible, rather than by individual matching at the design stage. Of course, it remains true that spatial effects will usually only be of scientific interest if they persist after adjustment for all known risk factors, whether or not the risk-factors themselves are spatially structured.

9.5.3 Is stratification or matching helpful?

In general, epidemiologists are divided on the merits of stratification and/or matching over the completely randomised design, for the good reason that there are clear advantages and disadvantages to the more complex designs and the balance between the different considerations will inevitably change in different practical settings. See, for example, Woodward (1999, pp. 266–8).

When spatial variation is a major scientific focus, the author's opinion is that fine stratification and individual matching are both undesirable, because they severely complicate the interpretation of of the estimated spatial effect. Specifically, consider the problem of estimating residual spatial variation in risk in a stratified design. As noted above, the analysis proceeds in the first instance by adding stratum effects to the generalized additive model (9.8). If we now let $Y_{ij} = 1/0$ denote whether the jth event in the ith stratum is a case or control, respectively, and write $p_{ij} = \mathrm{P}(Y_{ij} = 1)$, then the model for the data is that the Y_{ij} are mutually independent with

$$\log\{p_{ij}/(1 - p_{ij})\} = \alpha_i + z'_{ij}\beta + \ell(x_{ij}) \tag{9.16}$$

where x_{ij} and z_{ij} respectively denote the location and covariate vector associated with Y_{ij}. In the extreme case of individual matching, or more generally when there are many strata, the standard analysis uses the log-likelihood (9.15) to estimate β and $\ell(\cdot)$ whilst eliminating the nuisance parameters α_i. However, and especially when $\ell(\cdot)$ is specified non-parametrically, the interpretation of the resulting estimate of the spatial surface $\ell(x)$ is now problematic. Suppose, for example, that the events within a particular stratum are concentrated within a small sub-region of the whole study region. Then, the presence of the α_i parameter in (9.16), coupled with its elimination from the stratified log-likelihood (9.15) means in effect that the behaviour of the spatial surface $\ell(x)$ in that sub-region is only identifiable up to an arbitrary constant. More generally, one of the advantages claimed for stratified or matched designs is to eliminate the effects of the matching variables on the presumption that these are not of scientific interest; but if the matching variables are not spatially neutral, then they are partially confounded with the spatial effect, which for the purposes of the present discussion *is* of scientific interest.

9.6 Disentangling heterogeneity and clustering

An issue which arises quite generally in the analysis of spatial point process data, but which is particularly obvious in epidemiological applications, is the difficulty of separating variation in intensity from clustering of events. We have shown how the case-control paradigm can resolve the difficulty in the sense of enabling a test of a null hypothesis that specifies no variation in intensity *and* no clustering, but it leaves ambiguous the interpretation of a significant test result.

In other branches of statistics that deal with dependent data, for example in the analysis of real-valued spatial data, the usual pragmatic strategy is to partition the data into the sum of a spatially varying mean value function and random variation about the mean, hence $Y(x) = \mu(x) + Z(x)$, where $Y(\cdot)$ is the observed process, $\mu(x) = \mathrm{E}[Y(x)]$ and $Z(\cdot)$ is a zero-mean residual process (see, for example, Chiles and Delfiner, 1999). The analogous modelling assumption for point process data is to allow a non-constant intensity $\lambda(x)$ but to assume that the higher-order random variation is, in some sense, stationary. One way to formalise this, as discussed in Section 4.2, is to assume that the process is *re-weighted second-order stationary*, meaning that $\lambda_2(x,y)/\{\lambda(x)\lambda(y)\} = \rho(t)$, where t is the distance between x and y. For processes of this kind, the definition of the K-function extends naturally to

$$K_I(t) = 2\pi \int_0^t \rho(x)x\,dx$$

although, as discussed in Section (4.6.2) estimation of $K_I(t)$ from a single realisation is problematic.

In the case-control setting, there is a much clearer rationale for separate estimation of a spatially varying $\lambda(x)$ and stationary second-order properties. By construction, a random sample of controls constitute a realisation of an inhomogeneous Poisson process, albeit one with a possibly very complicated intensity function. The controls can therefore be used to estimate $\lambda(x)$. Given an estimate $\hat{\lambda}(x)$, we can then use the case data to estimate $\rho(t)$ or, equivalently, $K_I(t)$ incorporating the control-based estimate $\hat{\lambda}(x)$. Diggle *et al.* (2007) developed this idea and showed that a test for spatial clustering based on the empirical function $\hat{K}_I(t) - \pi t^2$ is competitive with existing approaches.

10

Spatio-temporal point processes

CONTENTS

10.1   Introduction

Until now, our focus has been on describing the distributions of points in two-dimensional space. We now extend the scope of our discussion to distributions in two-plus-one-dimensional space-time. Note that in this context, two plus one does not equal three, in the sense that the time dimension is fundamentally different from either of the two spatial dimensions.

Most spatial processes in nature are merely snapshots of evolving spatio-temporal processes. But to argue that they should therefore be analysed using spatio-temporal methods would be misguided. We analyse purely spatial data, and build purely spatial models, if and when in so doing we can address interesting scientific questions. By the same token, our aim for the remainder of this book is to describe statistical models and methods that can be used to analyse patterns of points in space-time when the questions of scientific interest concern both their spatial and their temporal behaviour, and cannot be answered by separate analyses of the spatial and temporal components of the spatio-temporal data.

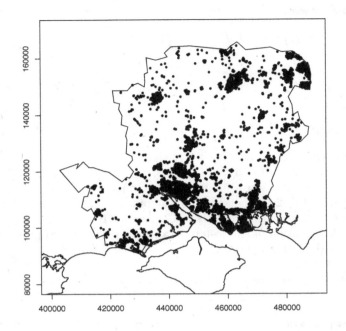

FIGURE 10.1
Locations (residential post-codes) of 10,572 successive cases of non-specific gastro-intestinal disease, as reported to NHS Direct in Hampshire, UK, between 1 January 2001 and 31 December 2003.

10.2   Motivating examples

10.2.1   Gastro-intestinal illness in Hampshire, UK

Figure 10.1 shows the locations of 10,572 cases of non-specific gastro-intestinal disease in the county of Hampshire, UK, as reported to NHS Direct (a phone-in triage service operating within the UK's National Health Service). Each location corresponds to the post-coded location of the residential address of the person making the call to NHS Direct. The data include all such reported cases between 1 January 2001 and 31 December 2003. A more informative display is an animation, an example of which can be viewed from the book's web-site.

The most obvious feature of these data is that the spatial pattern of calls predominantly reflects the spatial distribution of the underlying population; predominantly rather than exactly, for at least two reasons. Firstly, both dis-

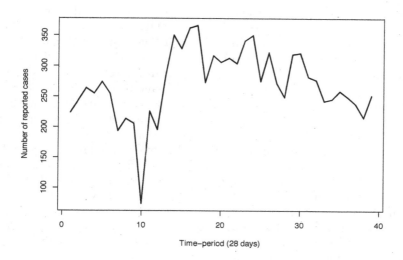

FIGURE 10.2
Times (days since 1 January 2001) of 10,572 successive cases of non-specific
gastro-intestinal disease, as reported to NHS Direct in Hampshire, UK, between 1 January 2001 and 31 December 2003.

ease risk and usage of NHS Direct vary between different demographic and
socio-economic groups within the population. Secondly, the diseases covered
by the blanket term "non-specific gastro-intestinal disease" include specific
conditions whose incidence is largely endemic but with occasional outbreaks,
typically associated with a contaminated food-source.

Figure 10.2 shows the numbers of reported cases per 28-day interval over
the three-year period covered by the data. The sharp dip in interval 10 is due
to a gap of 19 days between successive case-reports that is almost certainly
artificial. More generally, the relatively low numbers of reported cases in the
first of the three years reflects a progressive increase in the usage of what was
at the time a newly established service. The apparent decreasing trend over
the second and third years is harder to explain, but presumably includes a
combination of changes in case-incidence and in reporting behaviour.

The main objective in analysing these data is to develop a surveillance
system that would enable the timely identification of unusual spatially and
temporally localised peaks in incidence. Such unusual features, or *anomalies*,
within the overall pattern might point to an emerging public health problem;
see Diggle *et al.* (2003). We shall consider the data from this perspective in
Chapter 12.

10.2.2   The 2001 foot-and-mouth epidemic in Cumbria, UK

Figure 10.3 shows the culmination of the epidemic of foot-and-mouth disease (FMD) that affected many parts of the UK in 2001. The left-hand panel shows the locations of 5,153 animal-holding farms in the county of Cumbria at the start of the epidemic; the locations of the 658 amongst these that suffered cases of FMD are highlighted. Most of the county is intensively farmed, except for the mountainous area in the centre. The right-hand panel of Figure 10.3 gives the cumulative numbers of affected farms between January and July 2001. This shows the classic features of an epidemic curve, with an exponential-like early growth followed by a slowing in the rate of occurrence of new cases. For obvious reasons, the epidemic was not left to run its natural course, but was brought under control through a policy of aggressive control measures. These included slaughtering the stock of each infected farm, and of all other farms within a radius of a few kilometres, as soon as possible following the confirmation of a new case.

The epidemic began in the far north of the county and subsequently spread both south-west and south-east. Transmission of infection is thought to occur primarily between neighbouring farms but cases can also occur far from all pre-existing cases, possibly because of the unintended transport of infected material. As with our previous example an animation, again available from the book's web-site, is more informative. It shows that the final pattern of affected farms can be explained predominantly by the spread of infection from an initial case in the north-east corner of the county, together with a few smaller, apparently spontaneous outbreaks around locations relatively remote from all prevalent cases.

The scientific interest in analysing these data lies primarily in modelling the transmission of infection between farms, with a view to informing control strategies for future epidemics. We shall consider the data from this perspective in Chapter 13.

10.2.3   Bovine tuberculosis in Cornwall, UK

Figure 10.4 shows the locations of cattle farms in the county of Cornwall, UK, that tested positive for bovine tuberculosis (BTB) in an approximately annual series of inspections between 1989 and 2002. The original data-set also contained genotyping data, which we do not consider here; see Diggle, Zheng and Durr (2005).

In Figure 10.4, the grey-scale shading of each affected farm location corresponds to the year in which the farm first tested positive. Some farms tested positive in more than one year – the 919 cases are associated with 738 different farms.

One question of scientific interest is the extent to which the pattern of infected farms changes from year to year. Figure 10.5 shows the annual number of test-positive farms. Whilst the data themselves give no information on

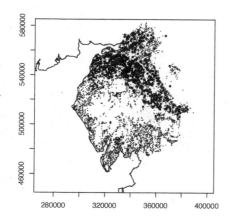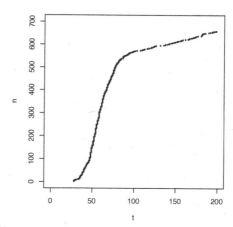

FIGURE 10.3
Data from the 2001 UK foot-and-mouth epidemic. Left panel: small dots show the locations of all animal-holding farms in the county of Cumbria, larger dots show the locations of those farms that experienced foot-and-mouth during the epidemic. Right panel: cumulative distribution of reporting times of cases, in days since 1 February 2001.

possible changes over time in the number of at-risk farms or the details of the testing regime, the rise in the number of test-positive farms from the mid-1990's is striking, and it is well-established that there has been a large increase in prevalence over the period in question; see, for example, Jalava *et al.* (2007).

10.3   A classification of spatio-temporal point patterns and processes

A *point* process is a stochastic process whose realisations consist of countable sets of points, which in this book we call *events*, in a pre-defined space. Correspondingly, a point pattern is a finite set of points that can usefully be treated as a partial realisation of a stochastic process; and by "usefully" we mean that treating the data in this way helps to answer an interesting scientific question.

In the spatial setting, we have always assumed that the space on which the events occur is a *continuous* region of the plane. In the extension to spatio-temporal processes and patterns, it is useful to allow either the spatial or temporal dimensions to be discrete.

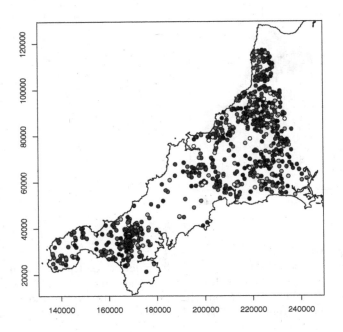

FIGURE 10.4
Locations of farms in Cornwall, UK, that tested positive for bovine tuberculosis at least once during the period 1989 to 2002. The grey-scale shading of the locations corresponds to the year in which each farm first tested positive (from white in 1989 to black in 2002).

The motivating examples in Section 10.2 cover the three cases of most interest. In each case, we can represent the data as the locations, x_i say, of the events of interest and the corresponding times, t_i say, at which they occur, hence $\{(x_i, t_i) : i = 1, ..., n\}$ where each $(x_i, t_i) \in A \times T$ for some pre-defined spatial region A and temporal region T.

In the example of Section 10.2.1 both the spatial and temporal regions are essentially continuous ("essentially" because of limited data-resolution). The region A is the county of Hampshire, T is the continuous time-interval between 1 January 2001 and 31 December 2003, and each reported case (x_i, t_i) could, in principle, have occurred at any place and time within $A \times T$.

In the example of Section 10.2.2, T is the continuous time-interval between 1 January and 31 July 2001, but A is now a discrete set, consisting of the locations of all stock-holding farms within the county of Cumbria as of 1 January 2001. Note, however, that by defining A in this way, we are making

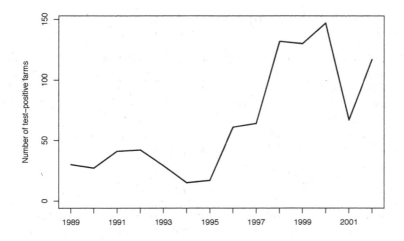

FIGURE 10.5
Number of farms in Cornwall, UK, testing positive for bovine tuberculosis in each year, 1989 to 2002.

the implicit assumption that the locations of the farms are not of any interest in themselves, otherwise we would have defined A to be the whole of Cumbria and farm locations as events. What *is* of interest is how the farms' locations collectively affect the progress of the epidemic. Note also that simply to refer to the times t_i at which particular farms reported FMD cases is only half the story. We need to attach to each farm both a time, t_i, and a label, or mark, that indicates into which of three categories the farm in question falls: newly reported cases at time t_i; stock pre-emptively culled at time t_i; no reported cases by the end of the observation period, in which case t_i corresponds to 31 July 2001.

In the example of Section 10.2.3, the spatial region A is continuous whilst the temporal region T is discrete, because incident cases are only identified annually.

We refer to these three situations as *continuous*, *spatially discrete* and *temporally discrete* spatio-temporal point processes, respectively.

10.4   Second-order properties

In this section, we define the first and second moment properties of a continuous, orderly spatio-temporal point process on $\mathbb{R}^2 \times \mathbb{R}^+$. The definitions involve essentially only notational changes from the analogous definitions for spatial point processes, as discussed in Chapter 4. Spatially discrete and temporally discrete processes can formally be treated as multivariate temporal and spatial point processes, respectively.

The first-order properties of a continuous spatio-temporal point process are described by its *spatio-temporal intensity function,*

$$\lambda(x,t) = \lim_{|dx|,|dt| \to 0} \left\{ \frac{E[N(dx,dt)]}{|dx||dt|} \right\}.$$

For a spatially or temporally stationary process, $\lambda(x,t)$ is independent of x or t, respectively. For a spatio-temporally stationary process, $\lambda(x,t)$ assumes a constant value λ, representing the mean number of events per unit area per unit time. Note that from a strict mathematical perspective it makes no sense to talk about the marginal spatial or temporal properties of a stationary spatio-temporal point process; in particular, both the mean number of events per unit area and the mean number per unit time are infinite. In practice, we only ever observe a spatio-temporal process on a finite region, $A \times T$, in which case we can always define marginal spatial and temporal intensities as, respectively,

$$\lambda_T(x) = \int_T \lambda(x,t)dt \quad \lambda_A(t) = \int_A \lambda(x,t)dx.$$

The same point applies to other summary properties, and we shall not labour it further, other than to emphasise the need for care when interpreting "marginal" spatial or temporal properties that depend on the observation region. Note also that because point process intensities are not normalised, we can interpret the spatio-temporal intensity $\lambda(x,t)$ in three different ways: as a joint spatio-temporal intensity; as a conditional spatial intensity for any given value of t; or as a conditional temporal intensity for any given value of x.

The *second-order spatio-temporal intensity function* is similarly defined as

$$\lambda_2(x,y,s,t) = \lim_{|dx|,|dy|,|ds|,|dt| \to 0} \left\{ \frac{E[N(dx \times ds)N(dy \times dt)]}{|dx||dy||ds||dt|} \right\},$$

and the *second-order conditional spatio-temporal intensity* as

$$\lambda_c(x,s|y,t) = \lambda_2(x,y,s,t)/\lambda(y,t),$$

corresponding to the intensity at (x, s) conditional on the information that there is an event at (y, t).

For a spatio-temporally stationary, isotropic process, $\lambda_2(x, s, y, t)$ reduces to $\lambda_2(u, v)$, where $u = ||x - y||$ and $v = |s - t|$. The function $\rho(u, v) = \lambda_2(u, v)/\lambda^2$ is then called the *spatio-temporal pair correlation function*. Note that we are here using "isotropic" as a convenient shorthand for "spatially isotropic and temporally reversible." Temporal reversibility is an innocuous assumption if we are concerned only with second-order properties but in general, and as we shall discuss in a later section, the directional nature of time can often be used to good effect in devising spatio-temporal models and associated methods of analysis.

The concept of *intensity re-weighted (second-order) stationarity* (Baddeley, Møller and Waagepetersen, 2000) also extends directly to the spatio-temporal setting. Provided that $\lambda(x, t)$ is bounded away from zero, intensity re-weighted (second-order) spatio-temporal stationarity requires that

$$\lambda_2(x, s, y, t)/\lambda(x, s)\lambda(y, t) = \rho(u, v) \tag{10.1}$$

depends only on $u = ||x - y||$ and $v = |s - t|$.

The K-function of a stationary, isotropic spatio-temporal point process can be defined as

$$K(u, v) = \lambda^{-1}\mathrm{E}[N_0(u, v)], \tag{10.2}$$

where $N_0(u, v)$ is the number of further events within distance u and time v of an arbitrary event. Two variants of this definition are available according to whether we consider the occurrence of events before and after, or only after, the arbitrary event. We will use the second of these, in which case the K-function of a homogeneous spatio-temporal Poisson process is

$$K(u, v) = \pi u^2 v : u \geq 0, v \geq 0.$$

Provided that the process is orderly, the link between $K(u, v)$ and $\lambda_2(u, v)$ is then that

$$\lambda K(u, v) = 2\pi\lambda^{-1} \int_0^v \int_0^u \lambda_2(x) x \, dx \, dt \tag{10.3}$$

For an intensity re-weighted stationary process, we therefore define an *inhomogeneous K-function* as

$$K_I(t) = 2\pi \int_0^v \int_0^u \rho(x, t) x \, dx \, dt$$

which reduces to (10.3) in the stationary case.

10.5   Conditioning on the past

The *history* of a spatio-temporal point process at time t is the collection of all events of the process, (x_i, t_i) say, that occur before time t. We write this as $\mathcal{H}_t = \{(x_i, t_i) : t_i < t\}$. The *complete conditional intensity* of the process is then defined as

$$\lambda_c(x, t | \mathcal{H}_t) = \lim_{|dx|, |dt| \to 0} \left\{ \frac{E[N(dx, dt) | \mathcal{H}_t]}{|dx||dt|} \right\}.$$

Informally, this describes how the likelihood of observing an event at location x and time t changes as the partial realisation of the process up to, but not including, time t, develops over time.

The complete conditional intensity provides a simple and intuitive way of defining a *spatio-temporal Poisson process* as a continuous, orderly process for which $\lambda_c(x, t) = \lambda(x, t)$, for all (x, t).

More generally, whilst the moment properties considered in Section 10.4 are all summary properties of a spatio-temporal point process, and in particular different processes can share the same summary descriptions, the complete conditional intensity characterises an orderly spatio-temporal point process uniquely. In later chapters, we will make use of this in developing models and associated methods of inference for spatio-temporal point process data. Here, we illustrate the point by considering a general algorithm for simulating any continuous, orderly process on a finite region $A \times T$, where T is the interval $[0, t_0]$. Denote the events of the realisation in their order of occurrence by $(x_i, t_i) : i = 1, 2, ...$, i.e. $t_i < t_{i+1}$ for all $i \geq 1$, and recall that $\mathcal{H}_t$ denotes the history at time t. Note that, conditional on $\mathcal{H}_t$, the probability that no event occurs in the time-interval $[t, t + v]$ is

$$P_t(v) = \exp\left(-\int_t^{t+v} \int_A \lambda_c(x, s | \mathcal{H}_t) dx ds \right)$$

To initiate the simulation algorithm we need to choose $\mathcal{H}_0$, the initial configuration of events at time $t = 0$. The algorithm then proceeds as follows.

1. Set $i = 0$ and $t = 0$

2. Draw $U \sim \mathrm{U}(0, 1)$

3. Find v such that $P_0(v) = U$

4. If $v > t_0$, the simulated realisation is empty, i.e. there are no events in $A \times T$, otherwise proceed to

5. Set $i = i + 1$ and $t_i = t + v$

6. Draw x_i from the distribution on A with pdf proportional to $\lambda_c(x, t_i | \mathcal{H}_{t_i})$

7. Draw $U \sim U(0, 1)$

8. Find v such that $P_{t_i}(v) = U$

9. If $t_i + v > t_0$, the realisation is the set $(x_j, t_j) : j = 1, ..., i$, otherwise return to step 6

Although the above algorithm is completely general it is not always useful. The complete intensity is only one of a number of ways to define a spatio-temporal point process, and if $\lambda_c(x, t|\mathcal{H}_t)$ is not specified directly, its evaluation may be intractable. Note also that the necessary and sufficient conditions for a function $\lambda_c(x, t|\mathcal{H}_{\sqcup})$ to be valid as the complete conditional intensity of a spatio-temporal point process on $\mathbb{R}^2 \times \mathbb{R}^+$ are that $\lambda_c(x, t|\mathcal{H}_{\sqcup})$ should be non-negative valued and integrable over any finite sub-region $A \times T$ of $\mathbb{R}^2 \times \mathbb{R}^+$ for all possible configurations $\mathcal{H}_t$ at any time t. It is not obvious how one might check that a given function satisfies the second of these conditions. If we are willing to consider only process on a finite region $A \times T$, a sufficient condition for validity is that the value of $\lambda_c(x, t|\mathcal{H}_{\sqcup})$ is non-negative and finite for all (x, t) in $A \times T$ and all possible configurations $\mathcal{H}_t$ for any t in T. Whether the resulting process provides a realistic model in any specific applied setting is another matter.

In Chapter 13 we will discuss how to define scientifically interesting models through their complete conditional intensities. Here, we give a simple illustration of a process whose partial realisations combine elements of spatial regularity and spatial aggregation.

We take A to be the unit square, T the time-interval $[0, 100]$, and set the initial configuration, $\mathcal{H}_0$, as a single event located at the centre of A. For all $t \geq 0$, we define the complete conditional intensity as follows. Firstly, let

$$h(u) = \begin{cases} 0 & : \quad u < \delta, \\ 2 - \delta/(u - \delta) & : \quad \delta \leq u < 2\delta, \\ 1 & : \quad u \geq 2\delta. \end{cases} \tag{10.4}$$

Now, suppose that n_t events $(x_i, t_i) : i = 1, 2, ..., n_t$ occur before time t and let $u_i(x)$ denote the distance between x_i and an arbitrary location x. Finally, define the complete conditional intensity to be the function

$$\lambda_c(x, t|\mathcal{H}_t) = \prod_{i=1}^{n_t} h\{u_i(x)\}. \tag{10.5}$$

In Section 6.8.1, we used the interaction function (10.4) as a cautionary example against using a pairwise interaction point process to model a spatial distribution that combines elements of inhibition and aggregation. There, we argued that the corresponding pairwise interaction point process was unsatisfactory as a model because the equivalent spatial birth-and-death process did not have a realistic equilibrium distribution. Whether this is equally objectionable in the spatio-temporal setting depends on the context. The spatio-temporal process defined by (10.4) and (10.5) evolves in what is arguably a

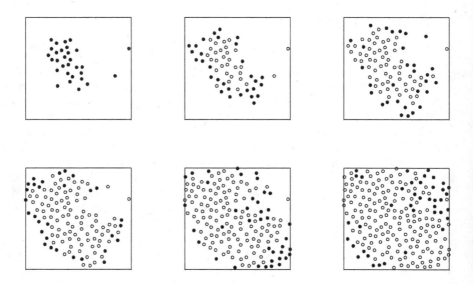

FIGURE 10.6
Snap-shots from a simulated partial realisation of a continuous spatio-temporal pairwise interaction point process with both inhibitory and aggregated characteristics. In each frame, the most recent 30 or 40 events are shown as solid dots, earlier events as open circles. See text for details of the simulation model.

sensible manner as time progresses. If we use (10.4) and (10.5) to define a spatio-temporal point process on a fixed region A, every realisation will terminate as soon as $\lambda_c(x,t|\mathcal{H}_t) = 0$ for all $x \in A$, at which point the process will be indistinguishable from a simple sequential inhibition process. However, the temporal progression of the process towards this terminating state does reflect both its inhibitory and its aggregated aspects. To illustrate this, Figure 10.6 shows a series of snap-shots of a realisation of the process, whilst Figure 10.7 shows a series of snap-shots of a purely inhibitory spatio-temporal process, again defined by (10.5), but now with a purely inhibitory interaction function,

$$h(u) = \left\{ \begin{array}{lll} 0 & : & u < \delta, \\ 1 & : & u \geq 0.05. \end{array} \right. \tag{10.6}$$

The contrast between the two processes as they develop over time is clear. More informatively, the book's web-site includes R functions that can be used to simulate repeated realisations of these and other processes, and to display these as animations.

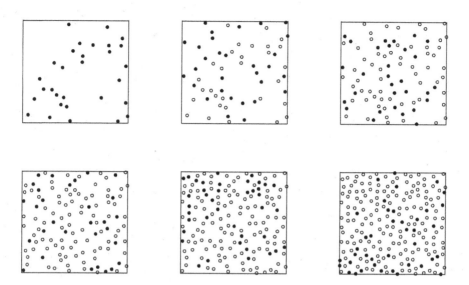

FIGURE 10.7
Snap-shots from a simulated partial realisation of a continuous spatio-temporal pairwise interaction point process with purely inhibitory characteristics. In each frame, the most recent 30 or 40 events are shown as solid dots, earlier events as open circles. See text for details of the simulation model.

10.6   Empirical and mechanistic models

The word 'model" means different things in different branches of science. In statistics, a model will always include a stochastic element, i.e. one or more random variables, and this could be taken as distinguishing statistical from (deterministic) mathematical models. However, a secondary and arguably more interesting distinction is between *empirical* models (whether statistical or mathematical), whose aim is only to describe the pattern of a data-set, and *mechanistic* models, which aim to encapsulate the underlying scientific processes that generated the data.

The author's opinion is that both kinds of models can be useful, and that the choice between an empirical and a mechanistic modelling strategy should be guided by both the context and the purpose of each application. For example, data collected in a controlled experimental environment are more likely to justify mechanistic modelling than are observational data. Chapters 12 and 13 cover empirical and mechanistic models, respectively.

11

Exploratory analysis

CONTENTS

11.1 Introduction

Throughout this chapter, unless otherwise stated we assume that the available data form a partial realisation of a spatially and temporally continuous process, consisting of the locations and times, $\{(x_i, t_i) : i = 1, ..., n\}$, of all the events of the process that lie within a designated spatio-temporal region $A \times T$. We also assume that the underlying process is orderly, so that ties amongst t_i arise only through round-off error and can legitimately be broken by randomly unrounding. Finally, without loss of generality we label the events in time-order so that, after unrounding if necessary, $t_i < t_{i+1}$ for all i.

In an exploratory analysis we may choose to treat the data *as if* the underlying process is as described above even when it is temporally or spatially discrete. For example, the gastro-intestinal illness data discussed in Section 10.2.1 strictly are confined to the finite set of Hampshire post-codes, each of which is simply a reference location within an area. However, the relatively fine spatial resolution of the UK post-code system (to within a single street in urban areas) justifies modelling the data as a point process in a spatial continuum. In contrast, for the foot-and-mouth epidemic data discussed in Section 10.2.2, events can only occur at farm locations, all of which are recorded whether or not they experience cases of foot-and-mouth, and any analysis of the data should acknowledge this. Finally, for the bovine tuberculosis data discussed in Section 10.2.3, the time-resolution of one year is sufficiently coarse that a sensible analytic strategy, at least for exploratory purposes, is to consider the data as arising from a discrete sequence of spatial point processes.

11.2   Animation

In Chapter 10 we described three motivating examples, and in so doing illustrated marginal graphical displays of the spatial and temporal dimensions of a spatio-temporal data-set. We also suggested that animations will often reveal rich structure that is hidden in marginal displays. The book's web-site includes a link to the R package stpp, which includes a simple animation function. More sophisticated forms of animation can also add considerable value. One such is to overlay the animation on a colour-coded contour map of a relevant spatially or spatio-temporally varying explanatory variable. Another is to add a slider that allows the user to control the speed (and direction) of the animation. The book's web-site includes links to several such examples.

11.3   Marginal and conditional summaries

Notwithstanding the above advice, the numerical and graphical summaries of spatial point processes described in the first part of this book can sometimes usefully be applied to spatio-temporal data in two different ways: marginally, by ignoring the time-dimension; and conditionally, by discretising the time-dimension and calculating spatial summaries within each time-interval so defined. The second of these is most likely to be helpful when the data are already coarsely discretised in time, as is the case for the bovine tuberculosis data introduced in Section 10.2.3.

11.3.1   Bovine tuberculosis in Cornwall, UK

The incidence of bovine tuberculosis (BTB) in the UK has increased markedly since the mid-1990's, with some of the highest rates occurring in the south-west of England. Figure 11.1 shows the annual numbers of herd-level incidences in the south-western county of Cornwall between 1996 and 2002, relative to the average annual numbers in the preceding seven years.

One of several questions posed by these data is whether the increase in incidence in the disease after 1995 applies equally across the whole county, or whether the spatial extent of the area at risk has changed since 1995. The available data refer only to farms that tested positive for the disease. Also, for reasons of confidentiality, in the version of the data that we analyse here and

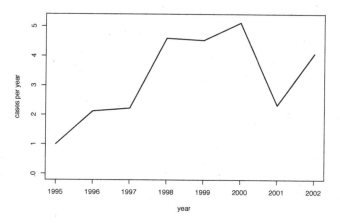

FIGURE 11.1

Annual numbers of incident bovine tuberculosis cases in Cornwall, relative to the average annual incidence over the baseline period 1989 to 1995.

that are available on the book's web-site, farm locations have been randomly jittered to preserve their anonymity.

For an exploratory analysis, we make the temporary working assumptions that the spatial distribution of at-risk farms has not changed over the period covered by the data, and that cases occur independently. Then, the process of case-locations in year t has intensity $\lambda_t(x) = \lambda(x)\rho_t(x)$, where $\lambda(x)$ represents the spatial intensity of at-risk farms whilst $\rho_t(x)$ represents spatial variation in risk. It follows that for any two time-periods t and s, $\lambda_t(x)/\lambda_s(x) = \rho_t(x)/\rho_s(x)$. In other words, we can estimate changes in the spatial distribution of risk over time by comparing intensities, exactly as described in Section 9.3. Because the annual numbers of cases were stable between 1989 and 1995, we treat the superposition of case-locations for these seven years as a baseline period, and compare the case-locations for each year $t > 1995$ with the baseline case-locations.

Let n_0 denote the number of baseline case-locations, $x_i : i = 1, ..., n_0$, counting each location more than once if the farm in question tests positive in more than one of the seven baseline years. For each year $t > 1995$, let n_t denote the number of year t case-locations. To compare year t with the baseline period, label the year t case-locations as $x_i : i = n_0 + 1, ..., n_0 + n_t$ and define binary random variables Y_{it} to have realised values let $y_{it} = 0 : i = 1, ..., n_0$, $y_{it} = 1 : i = n_0 + 1, ..., n_0 + n_t$. Then, conditional on the all $n_0 + n_t$ locations x_i, the Y_{it} are a set of independent Bernoulli trials with success probabilities $p_{it} = \rho_t(x_i)/\{\rho_0(x_i)+\rho_t(x_i)\}$, and we can estimate the corresponding spatially continuous surfaces $p_t(x)$ using the nonparametric binary regression method described in Section 9.3. Interpretation of the sequence of fitted binary regressions is easier if we use a common bandwidth for the kernel smoothing. Figure

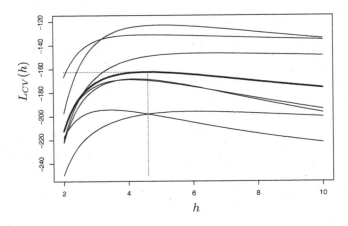

FIGURE 11.2

Cross-validated log-likelihood criteria, $L_{CV}(h)$, for kernel smoothing over bovine tuberculosis cases in Cornwall. Thin lines refer to comparisons between baseline period 1989 to 1995 and individual post-baseline years, whilst the thick line is calculated as the average of the $L_{CV}(h)$ over all seven post-baseline years. Dotted lines identify the location and value of the maximised average.

11.2 shows the cross-validated log-likelihood criterion (9.10), calculated separately for each year $t = 1996, \ldots, 2002$ and averaged over all seven post-baseline years.

Figures 11.3 and 11.4 show the resulting estimated incidence surfaces relative to the baseline period, using a common bandwidth $h = 4.6$km; note that the scale on the maps is in metres. In each case, a value of $p_t(x) > 1/7 \approx 0.14$ indicates an estimated risk for the year in question that is above the baseline risk. The maps show a number of interpretable features. Firstly, most of the estimates $p_t(x)$ are greater than 0.14, indicating that the increase in incidence has been widespread, rather than confined to local foci. Secondly, an area of relatively high incidence in the north-east of Cornwall, around the map reference (220000, 90000), increases both its numerical value and its spatial extent progressively between 1996 to 2000, falls in 20001 and rises again in 2002; recall from Figure 11.1 that the overall incidence also fell in 2001 but recovered in 2002. Thirdly, an area of relatively high incidence in the south-east in 1996, centered roughly on the map reference (190000, 50000), wanes in 1997, returns in 1998 and thereafter fluctuates in intensity from year to year.

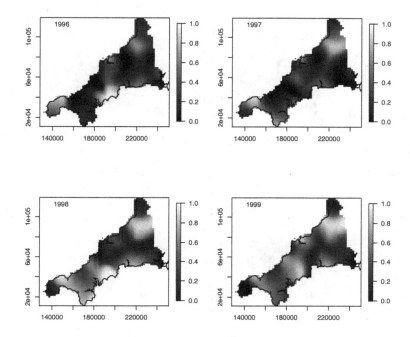

FIGURE 11.3

Kernel smoothing estimates of annual incidence of bovine tuberculosis cases in Cornwall, 1996 to 1999. In each case, the comparison is between the year indicated and the seven-year baseline period 1999 to 1995 and the common grey-scale runs from 0 (black) to 1 (white). The value $1/7 \approx 0.14$ corresponds to equality between current and baseline incidence.

11.4   Second-order properties

11.4.1   Stationary processes

Spatial and temporal units of measurement are fundamentally incompatible. It follows that overtly spatio-temporal versions of the kind of functional summaries discussed in earlier chapters in the purely spatial setting require at least two arguments. In this section, we consider non-parametric estimation of first-order and second-order structure.

The first-order structure of a stationary spatio-temporal point process is captured by a single parameter, λ, called the *intensity* of the process. The intensity is equal to the expected number of events per unit area per unit time. Hence, a natural estimator is the observed number of events per unit

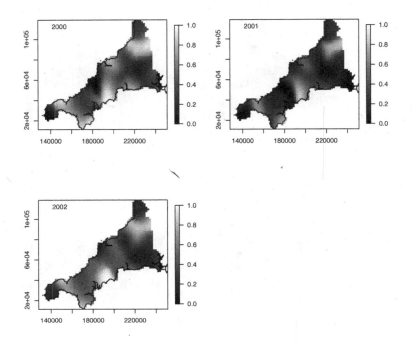

FIGURE 11.4

Kernel smoothing estimates of annual incidence of bovine tuberculosis cases in Cornwall, 2000 to 2002. In each case, the comparison is between the year indicated and the seven-year baseline period 1999 to 1995 and the common grey-scale runs from 0 (black) to 1 (white). The value $1/7 \approx 0.14$ corresponds to equality between current and baseline incidence.

area per unit time,

$$\hat{\lambda} = n/|A \times T|. \tag{11.1}$$

To estimate the second-order properties, we begin with the definition of $K(u, v)$ given as equation (10.2), replace the expected count on the right-hand-side by an observed count and include an edge-correction, analogous to the purely spatial case that was discussed in Chapter 4. This leads to the following estimator for $K(u, v)$. As in the purely spatial case, let $w(x, r)$ be the proportion of the circumference of the circle with centre x and radius r which lies within A, and write w_{ij} for $w(x_i, ||x_i - x_j||)$. Assume that the temporal observation window, T, is a simple interval, say $T = [a, b]$, and for any $v > 0$ let n_v denote the number of $t_i \leq b - v$. Then,

$$\hat{K}(u, v) = (nn_v)^{-1}|A|T \sum_{i=1}^{n_v} \sum_{j>i} w_{ij}^{-1} I(||x_i - x_j|| \leq u)I(t_j - t_i \leq v). \tag{11.2}$$

The estimate $\hat{K}(u, v)$ can most easily be examined as a grey-scale image or contour plot. For exploratory analysis, the plot can be assessed relative to either or both of the following benchmarks, according to their relevance to the application in hand. Firstly, for a homogeneous Poisson process,

$$K(u, v) = \pi u^2 v. \tag{11.3}$$

Secondly, for a process with independent spatial and temporal components,

$$K(u, v) = K_s(u)K_t(v) \tag{11.4}$$

where, with the proviso noted in 10.4, $K_s(u)$ and $K_t(v)$ are the marginal spatial and temporal K-functions.

To assess the significance of any departure from either of these benchmarks, we can use simple Monte Carlo methods as follows. To assess departure from the Poisson process, we compare $\hat{K}(u, v)$ with estimates calculated from independent simulations of a Poisson process conditioned to have n events in $A \times T$. To assess departure from independence, we compare with estimates calculated from independent permutations of the t_i holding the x_i fixed. In either case, for an exact test we need to specify a test statistic beforehand. The choice of test statistic should ideally take into consideration what would be a scientifically natural alternative hypothesis for each specific application. In the absence of any such natural alternative, a general-purpose statistic would be a summary measure of the discrepancy between $\hat{K}(u, v)$ and its theoretical form under the null hypothesis, for example

$$D = \int_0^{u_0} \int_0^{v_0} w(u, v)D(u, v)dudv, \tag{11.5}$$

where

$$D(u, v) = \{\hat{k}(u, v) - K_0(u, v)\}^2. \tag{11.6}$$

Implementation needs choices to be made for the truncation points u_0 and v_0, and for the weight function $w(u, v)$. Truncation points should depend both on the physical dimensions of $A \times T$ and on the context, whilst a natural choice for the weight function is $w(u, v) = (u^2v)^{-1}$.

An informal assessment of the fluctuations in $\hat{K}(u, v)$ relative to simulation envelopes is often at least as useful as a formal test, but their visualisation is less straightforward in the spatio-temporal setting than in the lower-dimensional purely spatial or temporal settings. One option is to superimpose a grey-scale image of $\hat{K}(u, v)$ and contour lines indicating regions where $\hat{K}(s, t)$ does or does not fall outside the relevant simulation envelope. Another is to add contour lines for p-values of local Monte Carlo tests, using test statistics $D(u, v)$ as defined at (11.6).

Rather than focus on $K(u, v)$, we could consider estimating the second-order intensity function or pair correlation function, $\lambda_2(u, v)$ or $\rho(u, v)$ respectively. The same considerations apply here as in the purely spatial case.

Firstly, estimating $\lambda_2(u, v)$ or other non-cumulative summaries requires the user to choose a smoothing constant, equivalent to choosing the bin-width in a histogram. Accordingly, the choice becomes less problematic as the number of events in the data-set increases. Secondly, some users may find it easier to interpret estimates of the non-cumulative functions $\lambda_2(u, v)$ and $\rho(u, v)$. There is, however, a simple connection between $\hat{K}(u, v)$ and a well-known test for space-time interaction proposed by Knox (1963, 1964). Knox's test consists of choosing values for u and v, summarising the data by a 2×2 contingency table that classifies each pair of events according to whether they are separated by less than or more than a distance u and time-interval v, and using as test statistic the number of pairs of events that are close both in space and in time. A Poisson approximation to the null distribution of the test statistic can be used provided u and v are sufficiently small that close pairs are rare, but an exact version for any values of u and v can be implemented as a Monte Carlo test by comparing the test statistic for the data with values obtained after random permutation of the event-times holding their locations fixed. Inspection of the right-hand side of (11.2) shows why the value of $\hat{K}(u, v)$ at any specific value of (u, v) can be interpreted as an edge-corrected version of Knox's statistic

11.4.2   Intensity-reweighted stationary processes

Recall that a point process is intensity-reweighted stationary if its intensity function, $\lambda(x, t)$, is bounded away from zero and its pair correlation function,

$$\rho(u, v) = \lambda_2(x, s, y, t) / \{\lambda(x, s)\lambda(y, t)\},$$

depends only on $u = ||x - y||$ and $v = |s - t|$, in which case its inhomogeneous K-function is

$$K_I(u, v) = 2\pi \int_0^v \int_0^u \rho(x, t)x \, dx \, dt.$$

Either non-parametric or parametric methods can be used to estimate $\lambda(x, t)$. However, as in the purely spatial case discussed in Chapter 4, it is difficult in practice to sustain an unambiguous distinction between first-order and second-order properties without making parametric assumptions. For non-parametric estimation of $\lambda(x, t)$, a pragmatic strategy is to make the working assumption that the first-order structure is separable, by which we mean that $\lambda(x, t)$ is a product of a function of x and a function of t, whilst second-order structure may be non-separable, by which we mean that $K_I(u, v)$ does not factorise into a product of a function of u and a function of v. To make the factorisation of $\lambda(x, t)$ unique, we need to impose a scaling condition. For data on $A \times T$, a convenient version is to write the intensity as $\lambda(x)\mu(t)$ where $\int_A \lambda(x)dx = 1$, in which case the mean number of events in any time-interval, (a, b) is $\int_a^b \lambda(t)dt$.

If we can specify a parametric model for the intensity function $\lambda(x, t)$, it can be estimated by maximum likelihood under the working assumption that the process is an inhomogeneous Poisson process. For example, we might specify $\log \lambda(x,t)$ as a linear regression model, with explanatory variables defined either as simple functions of x and t, i.e. a spatio-temporal trend surface, or more satisfactorily as scientifically relevant variables that can be measured throughout $A \times T$.

Once we have an estimate, $\hat{\lambda}(x,t)$ say, of the first-order structure, we can estimate $K_I(u,v)$ by an obvious modification of (11.2), namely

$$\hat{K}_I(u,v) = \frac{n}{n_v |A|T)} \sum_{i=1}^{n_v} \sum_{j>i} w_{ij}^{-1} I(||x_i - x_j|| \le u) \frac{I(t_j - t_i \le v)}{\hat{\lambda}(x_i,t_i)\hat{\lambda}(x_j,t_j)} \quad (11.7)$$

The form of the denominator on the right-hand-side of (11.7) suggests, correctly, that the estimator $\hat{K}_I(u,v)$ will be poorly behaved if the data span sub-regions where $\lambda(x,t)$ is close to zero, especially when we use non-parametric methods to estimate $\lambda(x,t)$. The same point applies in the purely spatial case, as discussed in Section 5.3.1, but tends to exacerbated in the spatio-temporal setting where the increase in the dimensionality of the study-region from two to three gives more scope for the occurrence of locally sparse data-configurations.

11.4.3 Campylobacteriosis in Lancashire, UK

Campylobacter is the most commonly identified cause of gastro-intestinal disease in the developed world. Temporal incidence of campylobacteriosis shows strong seasonal variation, rising sharply between spring and summer. Here, we describe the analysis of the locations and dates of notification of all reported cases of campylobacteriosis within the Preston postcode district (Lancashire, England) between January 1st 2000 and December 31st 2002. This analysis was previously reported in Diggle and Gabriel (2011).

The two panels of Figure 11.5 show the cumulative spatial distribution of the 969 reported cases and the cumulative distribution of the times, in days since 1 January 2000, on which the cases were reported. The spatial distribution of cases largely reflects the population at risk, consistent with the endemic character of gastro-intestinal infections, whilst the temporal distribution hints at a slight overall fall in the rate of incident cases over the three-year period, and some seasonal fluctuations.

These data can be considered as a single realisation of a spatio-temporal point process displaying a highly aggregated spatial distribution. As is common in epidemiological studies, the observed point pattern is spatially and temporally inhomogeneous, because the pattern of incidence of the disease reflects both the spatial distribution of the population at risk and systematic temporal variation in risk. When analysing such spatio-temporal point patterns, a natural starting point is to investigate the nature of any stochastic

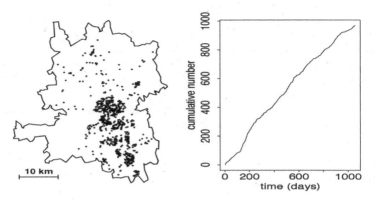

FIGURE 11.5

Campylobacteriosis in the district of Preston, 2000-2002. Locations of cases (left panel) cumulative distribution of reporting times in days since 1 January 2000 (right panel).

interactions amongst the points of the process after adjusting for spatial and temporal inhomogeneity.

The three panels of Figure 11.6 show: the study-region, corresponding to the Preston post-code sector of the county of Lancashire, UK; a grey-scale representation of the spatial variation in the population density, derived from the 2001 census; and the residential locations of the 619 recorded cases over the three years 2000 to 2002 in the most densely populated part of the study-region.

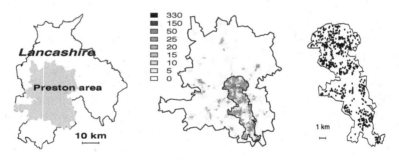

FIGURE 11.6

Lancashire campylobacteriosis data: study-area (left panel); 2001 population density in number of people per hectare (centre panel); locations of the 619 cases within the urban area (right panel).

We first estimate the marginal spatial and temporal intensities of the data. To estimate the spatial density, $\lambda(x)$, we use a Gaussian kernel estimator with band-width chosen to minimize the estimated mean-square error of $\hat{\lambda}(x)$, as suggested in Berman and Diggle (1989). To estimate the temporal intensity, $\mu(t)$, we use a Poisson log-linear regression model incorporating a time-trend, seasonal variation and day-of-the-week effects, hence

$$\log \mu(t) = \delta_{d(t)} + \sum_{k=1}^{3} \alpha_k \cos(k\omega t) + \beta_t \sin(k\omega t) + \gamma t,$$

where $\omega = 2\pi/365$ and $d(t)$ identifies the day of the week for day $t = 1, ..., 1096$. The sine-cosine terms corresponding to six-month and four-month frequencies are justified by likelihood ratio criteria under the assumed Poisson model, but this would over-state their significance if, as turns out to be the case, the data show significant spatio-temporal clustering. Figure 11.7 shows the resulting estimates of $\lambda(x)$ and $\mu(t)$.

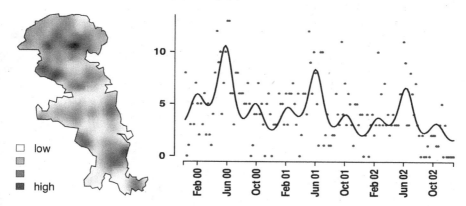

FIGURE 11.7
Lancashire campylobacteriosis data: kernel estimate of spatial intensity (left panel); weekly numbers of notified cases and fitted regression curve (right panel).

A comparison between the left panel of Figure 11.7 and the centre panel of Figure 11.6 shows, unsurprisingly, that cases tend to be concentrated in areas of high population density, whilst the right-hand panel of Figure 11.7 shows a decreasing time-trend and a sharp peak in intensity each spring. The smaller, secondary peaks in intensity are a by-product of fitting three pairs of sine-cosine terms and their substantive interpretation is open to question; here, we are using the log-linear model only to give a reasonably parsimonious estimate of the temporal intensity as a necessary prelude to investigating residual spatio-temporal structure in the data.

To investigate spatio-temporal structure, we consider the data in relation to two benchmark hypotheses. The hypothesis of *no spatio-temporal clustering*, H_0^C, states that the data are a realisation of an inhomogeneous Poisson process with intensity $\lambda(x)\mu(t)$. The hypothesis of *no spatio-temporal interaction, H_0^I*, states that the data are a realisation of a pair of independent spatial and temporal, re-weighted second-order stationary point processes with respective intensities $m(s)$ and $\mu(t)$. Note that in formulating our hypotheses in this way, we are making a pragmatic decision to interpret separable effects as first-order, and non-separable effects as second-order. Also, as here defined, absence of spatio-temporal clustering is a special case of absence of spatio-temporal interaction.

To test H_0^C, we compare the inhomogeneous spatio-temporal K-function of the data with tolerance envelopes constructed from simulations of a Poisson process with intensity $\hat{m}(s)\hat{\mu}(t)$. To test H_0^I, we proceed similarly, but with tolerance envelopes constructed by randomly re-labelling the locations of the cases holding their notification dates fixed, thus preserving the marginal spatial and temporal structure of the data without assuming that either is necessarily a Poisson process.

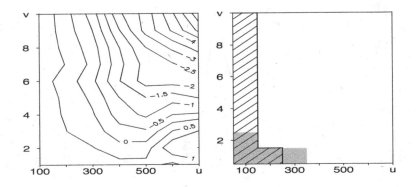

FIGURE 11.8
Lancashire campylobacteriosis data: $\hat{K}_{st}(u,v) - \pi u^2 v$, $\times 10^6$ (left panel); identification of sub-sets of (u,v)-space in which spatio-temporal clustering (diagonal black hatching) and/or spatio-temporal interaction (grey shading) is detected at the 5% level of significance (right panel).

The left-hand panel of Figure 11.8 shows $\hat{K}_{st}(u,v) - \pi u^2 v$ for the campylobacteriosis data. The diagonal black hatching on the right-hand panel of Figure 11.8 identifies those values of (u,v) for which the data-based estimate of $\hat{K}_{st}(u,v) - \pi u^2 v$ lies above the 95th percentile of estimated calculated from 1,000 simulations of an inhomogeneous Poisson process with intensity $\hat{\lambda}(x)\hat{\mu}(t)$. Similarly, the grey shading identifies those values of (u,v) for which $\hat{K}_{st}(u,v) - \hat{K}_s(u)\hat{K}_t(v)$ lies above the 95th percentile envelopes calculated from 1,000 random permutations of the x_i holding the t_i fixed.

The results suggest spatio-temporal clustering up to a distance of about 300 metres and a time-lag of 10 days, and spatio-temporal interaction at distances up to 400 metres and time-lags up to 3 days. These precise numbers should not be treated as anything other than rough estimates of the spatial and temporal scales on which the underlying disease process is operating. Nevertheless, they are broadly consistent with the infectious nature of the disease. Although the risk-factors for the disease are not completely understood, the time between infection and the emergence of symptoms is of the order of days rather than weeks, whilst spatial concentrations of cases are though to result multiple cases from a common source occurring relatively closely both in space and in time. The analysis also suggests the existence of stochastic structure that cannot be explained by the first-order intensity $\hat{\lambda}(x)\hat{\mu}(t)$. Note that the relatively large negative values of $\hat{K}_{st}(u, v) - \pi u^2 v$ at large values of u and v are not significantly different from zero, because the sampling variance of $\hat{K}_{st}(u, v)$ increases with u and v.

12

Empirical models and methods

CONTENTS

12.1 Introduction

Recall from Chapter 10 that the aim of empirical modelling is to describe the pattern in a data-set without necessarily pointing to any particular underlying mechanism. This approach can be useful in at least two different settings.

The first is when little is known about the underlying mechanism. A quantitative description of the pattern may then help to generate mechanistic hypotheses. The second is when the underlying mechanism is known to be sufficiently complex that formulating and fitting a mechanistic model is impractical, but also unnecessary to achieve the objectives of the analysis.

Suppose, for example, that we wish to analyse data on the locations and times of incident cases of a disease in a mixed urban/rural population, with a view to identifying anomalous patterns of incidence when and where they occur. The aetiology of the disease is only one of several factors that determine the observed pattern. A second, more obvious one is the geographical distribution of the population over the region of interest. Others may include socio-economic and behavioural risk-factors, patterns of travel between home and school or work-place, and the introduction of exotic infections by returning international travellers. An empirical model may then give useful predictions of where and when anomalies occur even if it cannot explain why.

12.2   Poisson processes

The definition of an inhomogeneous spatio-temporal Poisson process exactly parallels that of its spatial counterpart as defined in Section 6.4 except for the change from two to two-plus-one dimensions. Let AT denote any spatio-temporal region; typically in applications, AT will be of the form $S \times (0, T)$, for some spatial region S and time-interval $(0, T)$. Denote by $N(AT)$ for the number of events in AT. Then, an inhomogeneous spatio-temporal Poisson process with intensity $\lambda(x, t)$ is defined by the following two postulates:

ISTPP1 $N(AT)$ has a Poisson distribution with mean $\int_0^T \int_A \lambda(x, t) dx dt$.

ISTPP2 Given $N(AT) = n$, the n events in $A \times (0, T)$ form an independent random sample from the distribution on AT with pdf proportional to $\lambda(x, t)$.

Similarly, the associated log-likelihood function for data $(x_i, t_i) : i = 1, ..., n$ generated as a partial realisation of the process on AT is the direct spatio-temporal analogue of (8.1), namely

$$L(\lambda) = \sum_{i=1}^n \log \lambda(x_i, t_i) - \int_0^T \int_A \lambda(x, t) dx dt. \tag{12.1}$$

12.3   Cox processes

As discussed in Section 6.5, a Cox process is an inhomogeneous Poisson process whose intensity is itself a realisation of a non-negative-valued stochastic process. In the spatio-temporal setting, we write the intensity process as $\Lambda(x, t)$. The conditional Poisson property of the Cox process precludes any direct interactions between events. This makes it most appealing as a model when an observed pattern is thought to be determined by observed and/or unobserved environmental processes. If all of the relevant environmental variables are observed, an obvious provisional model would be an inhomogeneous Poisson process, with intensity $\lambda(x, t)$ specified as a regression on explanatory variables, $z(x, t)$ say. Note from the form of (12.1) that likelihood-based inference would strictly require all elements of $z(x, t)$ to be observed at all locations and times. If some of the relevant environmental variables are unobserved, we could represent them by stochastic processes, so defining a stochastic intensity $\Lambda(x, t)$. The model for the stochastic component of $\Lambda(x, t)$ could, in principle, be either mechanistic or empirical in character, but empirical models are more common in practice.

In what follows, we allow the expectation of $\Lambda(x, t)$ to vary with x and

t, but assume that its covariance structure is stationary. A convenient re-parameterisation is then to

$$\Lambda(x, t) = \lambda(x, t) R(x, t), \tag{12.2}$$

where $R(x, t)$ is a stationary process with expectation 1 and covariance function $\gamma(u, v) = \sigma^2 r(u, v)$. It follows that $\lambda(x, t)$ is the unconditional intensity of the point process, and the stationarity of $R(x, t)$ implies that the point process is intensity-reweighted stationary.

12.3.1 Separable and non-separable models

In the parameterisation (12.2), we say that $\Lambda(x, t)$ is *first-order separable* if

$$\lambda(x, t) = \rho(x) \mu(t) \tag{12.3}$$

and *second-order separable* if if

$$\gamma(u, v) = \sigma^2 r_1(u) r_2(v). \tag{12.4}$$

In most applications, it will be reasonable to assume that $R(x, t)$ is continuous at the origin, in which case so is $\gamma(u, v)$ and it follows that in (12.4), $r_1(u) = r(u, 0)$ and $r_2(v) = r(0, v)$.

To avoid ambiguity in (12.3), we scale the spatial intensity $\rho(x)$ so that $\int_A \rho(x) dx = 1$. This has the convenient consequence that $\mu(t)$ represents the expected number of events per unit time over the whole of the spatial region of interest, A. Assuming (12.3) is, in a sense, an arbitrary strategy, but can be defended pragmatically as follows. First-order and second-order effects cannot be distinguished empirically from a single realisation without additional assumptions. One way to maintain an operational distinction is therefore to treat spatially averaged time-trends and temporally averaged spatial trends as first-order, non-stochastic effects and any residual spatio-temporal structure as a second-order, stochastic effect.

The assumption of second-order separability assumption is also not particularly natural, but it is undeniably convenient. From the point of view of model formulation it is convenient that the product of any pair of valid spatial and temporal correlation functions is a valid spatio-temporal correlation function, hence the functions $r_1(u)$ and $r_2(v)$ can be chosen as any pair of valid correlation functions in $\mathbb{R}^2$ and $\mathbb{R}$, respectively. From the point of view of exploratory data analysis, separability makes for straightforward calculation and interpretation of non-parametric estimators for the correlation structure of $\Lambda(x, t)$.

The K-function of an intensity-reweighted stationary Cox process parameterised according to (12.2) is

$$K(u, v) = \pi u^2 v + 2\pi \lambda^{-2} \sigma^2 \int_0^v \int_0^u r(s, t) s \, ds \, dt, \tag{12.5}$$

where σ^2 is the variance of $R(x,t)$ and $r(s,t)$ its correlation function. Under separability, this simplifies to

$$K(u,v) = \pi u^2 v + 2\pi \lambda^{-2}\sigma^2 \int_0^u r_1(s)sds \int_0^v r_2(t)dt. \qquad (12.6)$$

Note that the standard estimator (11.2) for $K(u,v)$, neither assumes nor exploits second-order separability.

12.4 Log-Gaussian Cox processes

In a spatio-temporal log-Gaussian Cox process, $\log \Lambda(x,t)$ is a Gaussian process, and is therefore defined by its mean and covariance structure. In the intensity-reweighted stationary case with the parameterisation (12.2), we can write $R(x,t) = \exp\{S(x,t)\}$, where $S(x,t)$ is a Gaussian process with expectation $-0.5\nu^2$, variance ν^2 so that $E[R(x,t)] = 1$ as required, and correlation function $g(u)$. It follows that $R(x,t)$ has variance $\sigma^2 = \exp(\nu^2) - 1$ and correlation function $r(x,t) = \exp\{\nu^2 g(u,v)\} - 1$.

Self-evidently, any valid family of spatio-temporal correlation functions can be used to define a valid class of spatio-temporal log-Gaussian Cox processes. The study of such families of correlation function has generated a substantial literature in its own right, which is reviewed in Gneiting and Guttorp (2010).

As already noted, one way to guarantee validity is to use a separable family, $g(u,v) = g_1(u)g_2(v)$. For example, each of $g_1(u)$ and $g_2(v)$ could be chosen to lie within the Matérn class. This is convenient, and often provides a reasonable empirical fit to data, but is not especially natural from a mechanistic perspective.

An example of a physically motivated construction is given in Brown *et al.* (2000), who propose models based on a dispersion process. In discrete time, with δ denoting the time-separation between successive realisations of the spatial field, their model takes the form

$$S(x,t) = \int h_\delta(u)S(x-u, t-\delta)du + Z_\delta(x,t), \qquad (12.7)$$

where $h_\delta(\cdot)$ is a smoothing kernel and $Z_\delta(\cdot)$ is a noise process, in each case with parameters that depend on the value of δ in such a way as to give a consistent interpretation in the spatially continuous limit as $\delta \to 0$. Other parametric families of non-separable models are discussed in Cressie and Huang (1999), Gneiting (2002), Ma (2003, 2008) and Rodrigues and Diggle (2010).

12.5 Inference

As in the purely spatial case, in estimating the parameters of a spatio-temporal log-Gaussian Cox process, there is a choice between computationally easy but *ad hoc* moment-based methods and more demanding but principled likelihood-based methods. In principle, the theory described in Chapter 8 carries over directly, but at the time of writing I am not aware of any routine implementations in readily available software.

In epidemiological applications of spatio-temporal log-Gaussian Cox processes, a common scenario is that the events are cases of a particular disease, the intensity field $\Lambda(x, t)$ models spatio-temporal variation in disease risk and its deterministic (first-order) and stochastic (second-order) components represent, respectively, the effects of known and unknown factors that affect the observed pattern of cases. The former might typically include spatial variation population density, seasonal variation in population-averaged exposure to risk-factors and a variety of measured physical and social environmental variables that partially determine an individual's risk at a particular place or time. The latter would then represent a mix of unanticipated effects including unsuspected risk-factors, the temporary exposure of a sub-population to a spatially and temporally localised source of infection and infectious transmission between individuals. In the absence of a well-understood mechanistic model, this can lead to a focus on prediction rather than estimation. The practical goal is to uncover the behaviour of the unobserved realisation of $\Lambda(x, t)$, and in particular its stochastic component, either to provide clues about the aetiology of the disease in question or simply to identify anomalous departures from the normal pattern of disease incidence.

12.6 Gastro-intestinal illness in Hampshire, UK

We now describe an application of a log-Gaussian Cox process model to data on gastro-intestinal disease in the English county of Hampshire that were described in Section 10.2.1. The analysis was originally reported in Diggle, Rowlingson and Su (2005, henceforth DRS), who used only the 7126 cases reported in 2001 and 2002. As noted earlier, the goal of the analysis was to develop a method for identifying spatially and temporally localised peaks in incidence, termed *anomalies*. The analysis strategy assumed a log-Gaussian Cox process model with separable intensity and second-order separable latent Gaussian process. Hence,

$$\Lambda(x, t) = \rho(x)\mu(t) \exp\{S(x, t)\}, \tag{12.8}$$

where $S(x,t)$ has mean $-0.5\sigma^2$, variance σ^2 and correlation function $r(u,v) = r_1(u)r_2(v)$. Note, incidentally, that this does not equate to second-order separability of $\Lambda(x,t)$, whose correlation function is $\exp\{r_1(u)r_2(v) - 1\}$.

To model the time-trend in daily incidence, DRS made the working assumption that the daily counts of incident case-reports are independent, Poisson-distributed, with a log-linear specification for the mean, $\mu(t) = \exp\{\eta(t)\}$. The data show strong day-of-week effects, principally because of the differential use of the NHS Direct service between weekdays and weekends. Also, the general incidence in gastro-intestinal disease is known to vary seasonally, although the reasons for this are not fully understood. Finally, there may be long-term trends in incidence arising from the combined effects, again poorly understood, of a range of natural and social factors. To accommodate all of these features, DRS specified the model for η_t to be

$$\eta(t) = \alpha_{d_t)} + \beta t + \sum_{k=1}^{2}\{\gamma_k \cos(2k\pi t/365) + \delta_k \sin(2k\pi t/365)\}, \qquad (12.9)$$

where d_t denotes day-of-week. Figure 12.1 shows the observed and fitted daily incidence, in each case averaged over seven successive days to eliminate the day-of-week effects; note in particular the two seasonal peaks in incidence, in spring and late summer.

It is less clear how one might formulate a parametric model for the spatial variation in reported case incidence. This spatial variation must, to a considerable extent, reflect the spatial distribution of the underlying population, but is compounded with a variety of demographic and socio-economic factors that influence levels of usage of the NHS Direct service by different sub-populations. DRS used a non-parametric kernel smoothing approach. A standard kernel smoother of the kind described in Section 5.3 proved unsatisfactory because of the very wide variation in intensity between urban and rural locations. This resulted in many local estimates close to zero, which in turn led to unstable estimates of second-order properties as discussed below. DRS therefore used a kernel estimator with locally adaptive bandwidth. This takes the form

$$\hat{\rho}(x) = n^{-1}\sum_{i=1}^{n} h_i^{-2}k(||x - x_i||/h_i. \qquad (12.10)$$

In (12.10), h_i is intended to take relatively large values in areas where $\rho(x)$ is small, and *vice versa*. Following a recommendation in Silverman (1986), DRS specified h_i to be proportional to $\tilde{\rho}(x_i)^{-0.5}$, where $\tilde{\rho}(x)$ is a standard, fixed-bandwidth, kernel estimate. Figure 12.2 shows the resulting fitted surface $\hat{\rho}(x)$.

To estimate the second-order properties, DRS assumed a double-exponential model, $r(u,v) = \exp(-u/\phi_1)\exp(-v/\phi_2)$ and estimated the model parameters using a moment-based method suggested in Brix and Diggle (2001). This consisted of matching empirical and fitted spatial and temporal correlation functions according to a least squares criterion. Figure 12.3 shows

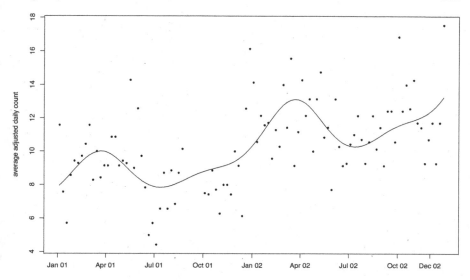

FIGURE 12.1
Gastro-intestinal illness in Hampshire: observed counts of reported cases per day, averaged over successive weekly periods (solid dots), compared with the fitted harmonic regression model of daily incidence (solid line).

the result. The *ad hoc* nature of the fitting process is open to criticism, but the fit appears reasonably good, and the ranges of the fitted spatial and temporal correlation are consistent with the presumption that clusters of nearby cases are most likely the result of sharing contaminated food, and with the acute, short-term nature of most food-borne infections.

To meet the objective of identifying anomalies, DRS followed Brix and Diggle (2001) in treating this as a problem in stochastic process prediction. Let $\mathcal{H}_t$ denote the dates and locations of all incident cases up to day t. Under the assumed model (12.8), let $R(x,t) = \exp\{S(x,t)\}$. Then, the formal solution to the prediction problem is the conditional distribution of $R(x,t)$ given $\mathcal{H}_t$. Similarly, the formal solution to the problem of forecasting anomalies with a lead-time of k days is the conditional distribution of $R(x,t+k)$ given $\mathcal{H}_t$. DRS used a plug-in version of the conditional distribution, meaning that they treated their estimates of $\rho(x)$, $\mu(t)$ and the parameters of $S(x,t)$ as if they were the true values. In general, a better and more elegant solution is to assign Bayesian priors to these unknown quantities; see, for example, Diggle *et al.* (2013). However, in the current application, parameter uncertainty makes only a small contribution to the overall predictive uncertainty, because all of the data inform the parameter estimates, whilst only local data inform the prediction of $R(x,t)$.

Figure 12.4 shows an example of the resulting predictive maps; more exam-

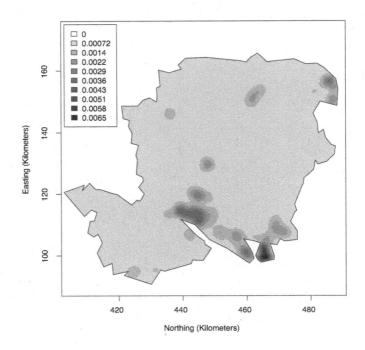

FIGURE 12.2
Gastro-intestinal illness in Hampshire: adaptive kernel estimate of the overall spatial variation in incidence, $\hat{\rho}(x)$.

ples can be inspected on the book's web-site. Note that what is being mapped is the predictive probability that $R(x,t)$ exceeds a specified value, c. The maps in Figure 12.4 is for $c = 2$. The author's view is that maps of this kind are more useful as an aid to decision-making than the more traditional practice of mapping the predicted value of $R(x,t)$ and its associated predictive standard deviation, because they focus attention on places and times at which there is a high probability of an important effect. An immediate corollary is that the decision-maker, rather than the statistician, should choose what value of c to use.

12.7 Concluding remarks: point processes and geostatistics

The log-Gaussian Cox process is a useful starting point for the analysis of spatio-temporal point patterns when the focus of scientific interest is on iden-

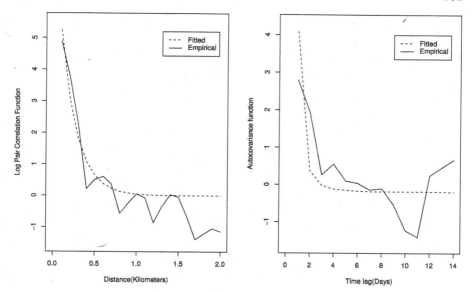

FIGURE 12.3

Gastro-intestinal illness in Hampshire: non-parametric (solid line) and fitted parametric (dashed line) log-transformed pair correlation functions (left-hand panel) and temporal covariance functions (right-hand panel).

tifying variation in the spatial and/or temporal intensity of events and both of the following conditions are satisfied: available explanatory variables do not completely explain the observed pattern of variation; and the underlying mechanism that generates the observed pattern is not well understood. The relative tractability of the process is also convenient for exploratory analysis by comparing empirical and theoretical moment properties.

There is a close connection between point process methods based on log-Gaussian Cox processes and geostatistical methods based on generalized linear mixed models (Diggle and Ribeiro, 2007). Geostatistical methods are concerned with the analysis of spatially discrete data relating to spatially continuous phenomena. In the spatio-temporal setting, a typical geostatistical data-set might consist of observed values y_i associated with spatio-temporal locations $(x_i, t_i) : i = 1, ..., n$, which are then linked to an observed spatio-temporal surface $u(x, t)$ and an unobserved stochastic process $S(x, t)$ by a suitable statistical model. As a specific example, suppose that conditional on $S(x, t)$, the y_i are realisations of mutually independent Poisson random variables with means

$$\mu_i = \exp\{\beta u(x_i, t_i) + S(x_i, t_i)\}. \tag{12.11}$$

Now suppose that the y_i in this hypothetical example were in fact counts of the numbers of event in small spatio-temporal regions AT_i centred on (x_i, t_i). Then, provided these small regions did not overlap, an assumed log-Gaussian

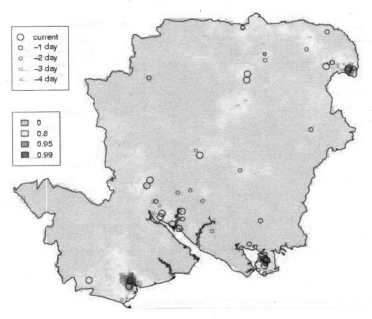

FIGURE 12.4
Gastro-intestinal illness in Hampshire: predictive map for 6 March 2003. The mapped value in each case is the conditional probability that $R(x,t) > 2$ given the data up to and including day t.

Cox process model for the underlying point process would again imply that conditional on $S(x,t)$, the y_i are realisations of mutually independent Poisson random variables, but now with means

$$\mu_i = \int_{AT_i} \exp\{\beta u(x,t) + S(x,t)\}dxdt. \tag{12.12}$$

Comparing (12.11) and (12.12), we see that the two models become equivalent if the AT_i are sufficiently small that both $u(x,t)$ and $S(x,t)$ can be treated as approximately constant within each AT_i.

A final comment is that whilst the log-transformation of the underlying Gaussian process $S(x,t)$ is convenient, it can induce very severe asymmetry in the distribution of $R(x,t)$, depending on the values of the mean and variance of $S(x,t)$. Within a Monte Carlo inferential framework, there is no particular reason to use the log-transformation: any non-negative-valued transformation could be used instead. Figure 12.5 shows, in a one-dimensional spatial setting, the effect of a range of transformations of the form $R(x) = a_k + b_k S(x)^{2k}$, where $S(x)$ is a realisation of a Gaussian process with mean 1, variance 1 and correlation function $\rho(u) = \exp(-u/0.1)$. The constants b_k have been chosen

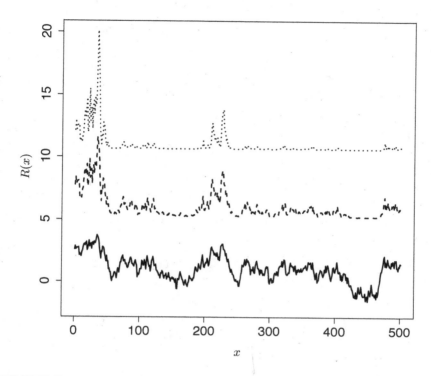

FIGURE 12.5
Three simulated transformed Gaussian processes, $R(x,t) = a_k + b_k S(x,t)^{2k}$, for $k = 0.5, 1, 2$ (solid, dashed and dotted lines, respectively).

so that $R(x)$ also has variance 1, whatever the value of k, whilst the a_k have been chosen simply to separate the three traces so as to ease the comparison.

13

Mechanistic models and methods

CONTENTS

13.1   Introduction

All natural processes evolve over time. It follows that a mechanistic model of a natural process must do likewise. This immediately switches the focus of statistical modelling and analysis from a joint description of spatial and temporal properties to a conditional description, in which the current properties of a process are specified conditionally on its realisation up to the current time, i.e. by conditioning on the past as described in Section 10.5.

13.2   Conditional intensity and likelihood

The *conditional intensity function* of a spatio-temporal point process, written $\lambda_c(x, t|\mathcal{H}_t)$, is the spatial first-order intensity at time t conditional on the *history*, $\mathcal{H}_t$ say, of the process up to time t, i.e. $\mathcal{H}_t$ is the set of locations and times of all events of the process that occur before time t. More formally, writing $N(dx, dt)$ for the number of events in an infinitesimal space-time region $ds \times dt$,

$$\lambda_c(x, t|\mathcal{H}_t) = \lim_{|dx| \to 0, dt \to 0} \left\{ \frac{\mathrm{E}[N(dx, dt)|\mathcal{H}_t]}{|dx|dt} \right\}. \qquad (13.1)$$

If the process is orderly, we can replace the expectation term on the right-hand-side of (13.1) by the probability that there is an event in the infinitesimal space-time region $dx \times dt$ plus terms that vanish in the limit, and the conditional intensity function completely defines the process.

Now consider the restriction of an orderly process to a finite but otherwise arbitrary spatial region A, and denote by $t + U$ and X the time and location, respectively, of the the first event that occurs after time t. Then,

$$P(U > u) = \exp\{-\int_t^{t+u} \int_A \lambda_c(s, t|\mathcal{H}_t) ds dt\}, \qquad (13.2)$$

and the conditional probability density of X given $U = u$ is proportional to the conditional intensity, $\lambda_c(x, t + u|\mathcal{H}_{t+u}) : x \in A$.

Together, these results show that any orderly spatio-temporal point process can be interpreted as a time-indexed sequence of inhomogeneous Poisson processes whose intensity evolves in response to its history. It follows that for data $\{(x_i, t_i) : i = 1, ..., n\}$ consisting of the locations and times of all events in a spatio-temporal $A \times [0, T]$, the log-likelihood is

$$L = \sum_{i=1}^{n} \log \lambda_c(x_i, t_i|\mathcal{H}_{t_i}) - \int_0^T \int_A \lambda_c(s, t) ds dt. \qquad (13.3)$$

Note that (13.3) is identical to (12.1) except that the conditional intensity, $\lambda_c(s, t|\mathcal{H}_t)$ in (13.3) replaces the unconditional intensity $\lambda(x, t)$ in (12.1).

An immediate consequence of (13.3) is that likelihood-based inference is, at least in principle, straightforward for any model that we choose to define by specifying its conditional intensity. Also, specifying a model in this way is scientifically appealing because of the direct relationship of the conditional intensity to an underlying mechanism.

Strictly, the only requirement for a valid specification of the conditional intensity is that $\lambda_c(x, t)$ is non-negative valued and integrable over A for any possible history at any time $t \leq T$. As a counterexample, consider the specification

$$\lambda_t(x) = \begin{cases} \lambda_0 & : \quad t = 0 \\ \theta^{n_t(x)} & : \quad t > 0 \end{cases}$$

where $\theta > 1$ and $n_t(x)$ is the number of events such that $t_i < t$ and $||x - x_i|| < \delta$, for some $\delta > 0$. The early development of this process resembles a homogeneous Poisson process, but as soon as a new event falls within distance δ of an existing event, this "seeds" a cluster that eventually grows uncontrollably. This is (not coincidentally) reminiscent of the unsatisfactory behaviour of a pairwise interaction point process with an attractive interaction function, as discussed in Section 6.8.1.

In order for the above results on the conditional intensity to be useful for inference, two further conditions need to hold. Firstly, unless a model is directly specified through its conditional intensity, we need to be able to derive an explicit expression for $\lambda_c(x, t)$, and this may be difficult, or impossible. For example, the conditional intensity of a log-Gaussian Cox process is intractable.

Secondly, the integrand for the integral term on the right-hand side of (13.3) is typically a complicated function of location and time with many local modes, making it difficult to evaluate the integral term accurately.

13.3   Partial likelihood

A useful variant of the likelihood (13.3), which greatly simplifies the associated computations, can be obtained when the process is spatially discrete, in the sense of the classification described in Section 10.3. The variant is a direct analogue of the partial likelihood introduced by Cox (1972b) in the context of proportional hazards modelling of survival data, and subsequently used in the point process setting by Møller and Sorensen (1998), Lawson and Leimich (2000) and Diggle (2006).

Denote by $\{(x_i, t_i) : i = 1, ..., n\}$ the observed events of the process in the spatio-temporal region $A \times (0, T)$, ordered so that $t_i < t_{i+1}$, and let $\{x_i : i = n + 1, ..., N\}$ be the set of locations of potential events that have not occurred by time T. For each of $i = 1, ..., n$, define

$$p_i = \lambda_c(x_i, t_i) / \sum_{j=i}^{N} \lambda_c(x_j, t_i). \tag{13.4}$$

Each p_i is the conditional probability that *the* event at time t_i is at location x_i, given that *an* event occurs at time t_i and at one of the locations x_j, $j \geq i$. It follows that

$$PL = \sum_{i=1}^{n} \log p_i \tag{13.5}$$

is the log-likelihood for the observed assignment of the times t_i to the locations x_i. Because each p_i involves a ratio of conditional intensities, it will typically only identify a sub-set of the model parameters. How much this matters depends on whether the unidentified parameters are of interest in their own right. In the original setting of survival analysis, the unidentified parameters correspond to an arbitrary baseline hazard function, $\lambda_0(t)$ say, and avoidance of the need to specify a parametric model for $\lambda_0(t)$ is sometimes advanced as a virtue of the method.

The computational advantage of (13.5) over (13.3) is that its evaluation requires only a finite summation, so no approximation is needed. Against this, the partial likelihood method is potentially inefficient, although the extent of its efficiency would seem to be context-specific.

Diggle, Kaimi and Abellana (2011) proposed an extension of the partial likelihood to spatially continuous processes. In this case, events could have occurred anywhere in A, and (13.4) is replaced by

$$p_i^* = \lambda_c(x_i, t_i) / \int_A \lambda_c(x, t_i) dx \qquad (13.6)$$

The need to evaluate the integral in the denominator of (13.6) loses some of the computational savings of the partial likelihood over the full likelihood (13.3).

13.4   The 2001 foot-and-mouth epidemic in Cumbria, UK

The analysis reported here is based on the account in Diggle (2006). We describe a spatial SIR (Susceptible, Infectious, Removed) model, similar to a model proposed by Keeling *et al.* (2001), and show how the partial likelihood method can give a computationally routine implementation of likelihood-based methods of inference within this class of models. We analyse the data introduced in Section 10.2.2 concerning the evolution of the 2001 FMD epidemic in Cumbria, the English county most severely affected by the epidemic.

The basis of the Keeling *et al.* (2001) model is a decomposition of the rate of transmission of the infection from an infectious farm i to a susceptible farm j into five terms, representing: a baseline rate; the infectiousness of the transmitting farm; the susceptibility of the receiving farm; the spatial juxtaposition of the transmitting and receiving pair; an at-risk indicator $I_{ij}(t)$. Hence,

$$\lambda_{ij}(t) = \lambda_0(t) A_j(t) B_i(t) C_{ij}(t) I_{ij}(t). \qquad (13.7)$$

The baseline hazard, $\lambda_0(t)$ is not identifiable from the partial likelihood, and we therefore leave its form unspecified. We assume that the terms A_i and B_j can each be described by regressions involving a vector of explanatory variables attached to each farm. Candidates for the analysis reported here are the numbers of cows, n_{1i}, and sheep, n_{2i}, held on farm i at the start of the epidemic, and the area of land owned by the farm, a_i say. Our regression models take the form

$$A_i = (\alpha n_{1i}^\gamma + n_{2i}^\gamma) \exp(a_i \delta), \qquad (13.8)$$

with a similar expression for B_i but replacing α by β. The rationale for this form of dependence on animal numbers was that linear dependence is a natural starting point, but there was a specific interest in establishing whether this was in fact the case. The rationale for including farm area as a multiplicative effect was that in a large farm, only animals relatively close to the farm's boundary would be likely to transmit infection to, or receive infection from, a neighbouring farm, thereby reducing the effective animal numbers.

For the spatial term C_{ij}, we write $d_{ij} = ||x_i - x_j||$ for the distance between farms i and j, and assume that

$$C_{ij} = \exp\{-(d_{ij}/\phi)^\kappa\} + \rho. \tag{13.9}$$

In (13.9) ϕ has an immediate interpretation as the rate at which the transmission of infection decays with increasing distance, whilst κ is a shape parameter whose value we fix at 0.5 to capture the sharper-than-exponential decay noted by Keeling *et al.* (2001). The parameter $\rho > 0$ allows for the occasional occurrence of new, spatially isolated cases which would otherwise distort the fit of the model to the prevailing pattern of transmission between near-neighbouring farms.

An important feature of the FMD epidemic is that both reactive and preemptive culling strategies were used to try to limit the spread of the epidemic. As soon as practicable after a farm was found to be infected, all of its animals and those of all other farms either close to, or otherwise considered to have been in dangerous contact with, the infected farm, were slaughtered. To capture this feature, we define a risk-set $\mathcal{R}_i$ to consist of all farms that have neither been reported as infected nor culled at time t_i. Hence, the at-risk indicator $I_{ij}(t)$ is defined to be 1 if farm i is infected and has not had its animals slaughtered by time t and farm j is not infected and has not had its animals slaughtered by time t. Culling dates and reporting dates of new infections are known exactly, but infection dates themselves are not. We make the simplifying assumption that each infection took place five days before the corresponding reporting date.

With the above definitions in place, we can evaluate the partial likelihood for the model as follows. The rate at which a susceptible farm, j say, at time t becomes infected is $\lambda_j(t) = \sum_i \lambda_{ij}(t)$. The contribution of the jth infection event to the partial likelihood is then $p_j = \lambda_j(t_j)/\sum_k \lambda_k(t_j)$. Maximisation of the partial likelihood with and without the farm-area term, $\exp(a_i\delta)$ on the right-hand side of (13.8) and the corresponding expression for B_i suggested that this term could be excluded, albeit not unequivocally so; the partial likelihood ratio statistic was 3.26 on 1 degree of freedom, corresponding to a p-value of 0.07. Table 13.4 gives the parameter estimates for the model without the farm-area effect.

The quoted confidence intervals in Table 13.4 are derived from a numerical estimate of the Hessian matrix, and their accuracy is suspect. A more reliable way to assess precision of estimation is through simulation, as described in Section 13.5.

The fitted spatial kernel, $f(d) = \exp\{-(d/\hat{\phi})^{0.5}\} + \hat{\rho}$, shows how the risk of transmission from an infected to a susceptible herd decays over an effective range of a few kilometres, as shown in Figure 13.4. This is consistent with the "ring culling" policy implemented during the epidemic, whereby any farm within 3km of an infected farm had its stock slaughtered to limit the risk of further transmission (Tildesley *et al.*, 2009). The estimate of γ indicates a sub-linear dependence of risk on animal numbers. One possible interpretation

TABLE 13.1
Parameter estimation for the five-parameter model fitted to combined data
from Cumbria and Devon

Parameter	Estimate	95% confidence interval	
α	1.42	1.13	1.78
β	36.17	0.19	692.92
γ	0.13	0.09	0.21
ϕ	0.41	0.36	0.48
ρ	1.3×10^{-4}	8.5×10^{-5}	2.1×10^{-4}

of this is that farms with larger animal holdings also occupy greater areas, and
transmission of infection occurs predominantly at or near boundaries between
farms. This could also explain why farm area in itself does not give a significant
improvement in the fit of the model after taking account of stocking numbers.

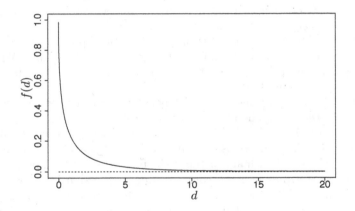

FIGURE 13.1
Estimated transmission kernel in the spatial SIR model fitted to data on the
2001 FMD epidemic in Cumbria, UK.

13.5 Nesting patterns of Arctic terns

This example is taken from a study of the nesting behaviour of Arctic terns
conducted in the Ebro Delta Natural Park, of Spain (Hernández and Ruiz,
2003). The analysis reported here is based on the account in Diggle, Kaimi

and Abellana (2010). The data consist of the locations x_i and times t_i of nests made by successive arrivals at a nesting colony on each of several small islets. Figure 13.5 shows the locations of the two largest colonies at the end of the nesting season. On each islet, height above sea-level was measured at each nest location and at the additional locations shown on each map. In what follows, we analyse only the larger of the two colonies, containing 104 nests.

Islet 84 **Islet 23**

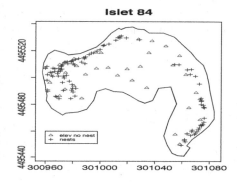
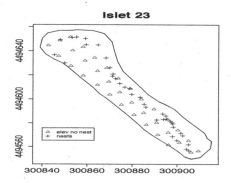

FIGURE 13.2
Locations of Arctic tern nests (crosses) on each of two islets in the Abro Delta National Park, Spain, and additional locations at which height above sea-level was also recorded (triangles).

Two features of the resulting point process that are not of direct interest but must be incorporated in any model of the process are the physical size of each nest and the birds' clear preference for locating their nests close to the shoreline, but not so close as to be at risk of inundation at high tide. The question of scientific interest is whether, after taking account of these features, incoming birds' choices of nesting sites show evidence of aggregation or inhibition, reflecting a preference to co-locate with established nests or territorial behaviour, respectively. We therefore model the conditional intensity as the product of three terms: a baseline intensity $\lambda_0(t)$ which we leave unspecified, a regression function $a(x)$ and a behavioural function $b(x,t)$.

For the regression function, we assume that

$$a(x) = \exp\{\alpha_1 z(x) + \alpha_2 z(x)^2\}, \qquad (13.10)$$

where $z(x)$ denotes the height above sea-level of the location x, and the quadratic term accommodates an anticipated non-monotone relationship.

To capture the behavioural aspects of the colonisation process, we first define a family of interaction functions,

$$h(d) = \begin{cases} 0 & : \quad d < \delta \\ 1 + \beta \exp\{-(u - \delta)^c\} & : \quad d \geq \delta \end{cases} \qquad (13.11)$$

In (13.11), δ represents the diameter of a typical nest, whilst α allows for either aggregative ($\beta > 0$) or inhibitory ($\beta < 0$) behaviour. The constant c is a shape parameter, which is difficult to estimate. We therefore consider only two candidate values, $c = 1$ or 2.

For the behavioural term $b(x, t)$, at each time t we allow an incoming bird's preference for establishing a nest at location x to depend either on all existing nest-locations, or only on the nest-location closest to x. Let $N(t)$ denote the number of nests already in place immediately before time t. If we allow all pre-existing next locations to affect the incoming bird's preferences, then

$$b(x, t) = \prod_{j=1}^{N(t)} h(||x - x_j||). \tag{13.12}$$

If we assume that only the location of the most recent arrival is relevant, then

$$b(x, t) = h(d_0(t)), \tag{13.13}$$

where $d_0(t) = \min_{j=1}^{N(t)} ||x - x_j||$ is the smallest of the $N(t)$ distances between x and each of the pre-existing nests.

In contrast to the foot-and-mouth example described in Section 13.4, the process of nest colonisation is spatially continuous, and evaluation of the partial likelihood requires numerical integration of the conditional intensity over the whole island. This in turn requires the height above sea-level, $z(x)$, to be available at every location x. The data record the values of $z(x)$ only at nest-locations and at a further set of reference locations. To obtain a complete surface of values of $z(x)$, we use a simple piece-wise constant interpolation on the Dirichlet tessellation of the locations at which $z(x)$ has been recorded, including locations on the shore-line where $z(x) = 0$.

The models with behavioural terms defined by (13.12) and by (13.13) are not nested, nor are the versions of each model with $c = 1$ and 2 nested, and formal likelihood ratio tests are not available. Nevertheless, comparison of maximised partial likelihoods favours (13.13) with $c = 1$. Our preferred model for the conditional intensity is therefore

$$\begin{aligned} \lambda_c(x, t | \mathcal{H}_t) &= \lambda_0(t) \exp\{\alpha_1 z(x) + \alpha_2 z(x)^2\} \\ &\times \ [1 + \beta \exp\{-(d_0(t) - \delta)/\phi\}] \times \mathrm{I}[d_0(t) > \delta], \end{aligned} \tag{13.14}$$

where, as earlier, $d_0(t) = \min_{j=1}^{N(t)} ||x - x_j||$. We estimate the non-regular parameter δ as the smallest observed distance between any two nests in the final pattern, hence $\hat{\delta} = 0.24$ metres. Maximum partial likelihood estimates and Monte Carlo standard errors for the remaining parameters are shown in Table 13.2.

The estimates of β and ϕ suggest aggregative social behaviour operating up to a scale of several metres. The regression parameter estimates $\hat{\alpha}_1 = 0.22$ and $\hat{\alpha}_2 = -0.0046$ suggest a unimodal effect of height above sea-level, with the

TABLE 13.2
Maximum partial likelihood estimates and standard errors for the model fitted to the Arctic tern nesting data. Standard errors are Monte Carlo approximations, calculated by re-fitting the model to 100 simulated data-sets.

Parameter	Estimate	SE
α_1	0.22	0.0039
α_2	-0.0046	9.5×10^{-5}
β	18.94	0.39
ϕ	2.97	0.56

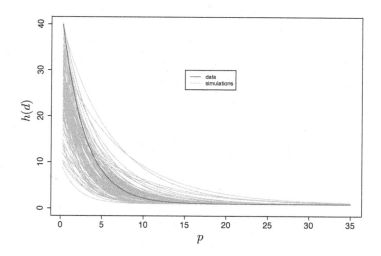

FIGURE 13.3
Estimated interaction function, $h(d)$, for the model fitted to the Arctic tern nesting data (thick line) and re-estimates from 100 simulated realisations of the model (thin grey lines).

most favoured locations at a height of $0.22/0.0092 \approx 24$ metres. The standard errors shown in Table 13.2 do not in themselves indicate with what precision we have been able to estimate the interaction function. Instead, we show this graphically in Figure 13.3, which compares the fitted interaction function $h(d)$ with 100 Monte Carlo re-estimates. This confirms that the aggregative behaviour is statistically significant, whilst indicating that both the strength of this effect and the effective range over which it operates are rather imprecisely estimated.

References

Alexander, F.E. and Boyle, P. (1996). *Methods for Investigating Localized Clustering of Disease.* Lyon : International Agency for Research on Cancer.

Anderson, N.H. and Titterington, D.M. (1997). Some methods for investigating spatial clustering, with epidemiological applications. *Journal of the Royal Statistical Society*, A **160**, 87–105.

Ardern, K. (2001). *Report on the Possible Increase in Cancer Cases in North Liverpool and Potential Links to the Site of the Former Incinerator at Fazakerley Hospital.* Liverpool : Liverpool Health Authority.

Baddeley, A.J. (1999). Spatial sampling and censoring. In *Stochastic Geometry: Likelihood and Computation*, eds. O.E. Barndorff-Nielsen, W.S. Kendall and M.N.M. Van Lieshout, 37–78. London : Chapman and Hall.

Baddeley, A.J. and van Lieshout, M.N.M. (1995). Area-interaction point processes. *Annals of the Institute of Statistical Mathematics*, **47**, 601–19.

Baddeley, A.J. and Møller, J. (1989). Nearest-neighbour Markov point processes and random sets. *International Statistical Review*, **57**, 89–121.

Baddeley, A.J., Møller, J. and Waagepetersen, R. (2000). Non- and semi-parametric estimation of interaction in inhomogeneous point patterns. *Statistica Neerlandica*, **54**, 329–50.

Baddeley, A.J., Moyeed, R.A., Howard, C.V. and Boyde, A. (1993). Analysis of a three-dimensional point pattern with replication. *Applied Statistics*, **42**, 641–68.

Baddeley, A.J. and Silverman, B.S. (1984). A cautionary example on the use of second-order methods for analyzing point patterns. *Biometrics*, **40**, 1089–93.

Baddeley, A. and Turner, R. (2000). Practical maximum pseudolikelihood for spatial point patterns (with Discussion). *Australian and New Zealand Journal of Statistics*, **42**, 283–322.

Bagchi, R., Henrys, P.A., Brown, P.E., Burslem, D.F.R.P., Diggle, P.J., Gunatilleke, C.V.C., Gunatilleke, I.A.U., Kassim, A.R., Law, R. and Valencia, R.L. (2011). Spatial patterns reveal negative density dependence and habitat associations in tropical trees. *Ecology*, **92**, 1723-1729

Bailey, T.C. and Gatrell, A.G. (1995). *Interactive Spatial Data Analysis*. Harlow : Longman.

Barnard, G. A. (1963). Contribution to the discussion of Professor Bartlett's paper. *Journal of the Royal Statistical Society*. B **25**, 294.

Bartlett, M. S. (1937). Properties of sufficiency and statistical tests. *Proceedings of the Royal Society*, A **160**, 268–282.

Bartlett, M. S. (1964). Spectral analysis of two-dimensional point processes. *Biometrika*, **51**, 299–311.

Bartlett, M. S. (1971). Two-dimensional nearest neighbour systems and their ecological applications. In *Statistical Ecology*, Vol. 1, eds. G.P. Patil, E.C. Pielou and W.E. Waters, 179–84. University Park: Pennsylvania State University Press.

Bartlett, M. S. (1975). *The Statistical Analysis of Spatial Pattern*. London: Chapman and Hall.

Bell, M.L. and Grunwald, G.K. (2004). Mixed models for the analysis of replicated spatial point patterns. *Biostatistics*, **5**, 633–648.

Berman, M. and Diggle, P. (1989). Estimating weighted integrals of the second-order intensity of a spatial point process. *Journal of the Royal Statistical Society*, B **51**, 81–92.

Berman, M. and Turner, T.R. (1992). Approximating point process likelihoods with GLIM. *Applied Statistics*, **41**, 31–8.

Bernal, J. D. (1960). Geometry of the structure of monatomic liquids. *Nature*. **185**, 68–70.

Besag, J. (1974). Spatial interaction and the statistical analysis of lattice systems (with discussion). *Journal of the Royal Statistical Society*, B **34**, 192–236.

Besag, J. (1975). Statistical analysis of non-lattice data. *The Statistician*, **24**, 179–95.

Besag, J. (1977). Contribution to the discussion of Dr Ripley's paper. *Journal of the Royal Statistical Society*, B **39**, 193–5.

Besag, J. (1978). Some methods of statistical analysis for spatial data. *Bulletin of the International Statistical Institute*, **47**, 77–92.

Besag, J. and Diggle, P. J. (1977). Simple Monte Carlo tests for spatial pattern. *Applied Statistics*, **26**, 327–33.

Besag, J. and Gleaves, J. T. (1973). On the detection of spatial pattern in plant communities. *Bulletin of the International Statistical Institure*, **45** (1), 153–8.

Besag, J., Milne, R. and Zachary, S. (1982). Point process limits of lattice processes. *Journal of Applied Probability*, **19**, 210–6.

Besag, J. and Newell, J. (1991). The detection of clusters of rare diseases. *Journal of the Royal Statistical Society*, A **154**, 143–55.

Bivand, R.S., Pebesma, E.J. and Gomez-Rubio, V. (2008). *Applied Spatial Data Analysis with R*. New York: Springer

Bostoen, K., Chalabi, Z. and Grais, R.F. (2007). Optimisation of the T-square sampling method to estimate population sizes. *Emerging Themes in Epidemiology*, **4**, doi:10.1186/1742-7622-4-7.

Breslow, N.E. and Day, N.E. (1980). *Statistical Methods in Cancer Research, Volume 1: The Analysis of Case-Control Studies*. Lyon : International Agency for Research on Cancer.

Brown, D. (1975). A test of randomness of nest spacing. *Wildfowl* **26**, 102-3.

Brown, D. and Rothery, P. (1978). Randomness and local regularity of points in a plane. *Biometrika*, **65**, 115–22.

Brown, P.E., Karesen, K., Roberts, G.O. and Tonellato, S. (2000). Blur-generated non-separable space-time models. *Journal of the Royal Statistical Society*, B **62**, 847–60.

Brown, S. and Holgate, P. (1974). The thinned plantation. *Biometrika*, **61**, 253–62.

Brown, T. (1979). Position dependent and stochastic thinning of point processes. *Stochastic Processes and their Applications*, **9**, 189–93.

Brown, T. C., Silverman, B. W. and Milne, R.K. (1981). A class of two-type point processes. *Z. Wahrscheinlichkeitstheorie verw. Geb.* **58**, 299–308.

Byth, K. (1982). On robust distance-based intensity estimators. *Biometrics* **38**, 127–35.

Byth, K, and Ripley, B. D. (1980). On sampling spatial patterns by distance methods. *Biometrics* **36**, 279–284.

Catana, A. J. (1963). The wandering quarter method of estimating population density. *Ecology* **44**, 349–60.

Chetwynd, A.G. and Diggle, P.J. (1998). On estimating the reduced second moment measure of a stationary spatial point process. *Australian and New Zealand Journal of Statistics*, **40**, 11–5.

Chetwynd, A.G., Diggle, P.J., Marshall, A. and Parslow, R. (2001). Investigation of spatial clustering from individually matched case-control studies. *Biostatistics*, **2**, 277–93

Chilès, J-P and Delfiner, P. (1999). *Geostatistics*. Hoboken: Wiley.

Chilès, J-P and Delfiner, P. (2012). *Geostatistics*, 2nd edn. Hoboken: Wiley.

Clark, P. J. and Evans, F. C. (1954). Distance to nearest neighbour as a measure of spatial relationships in populations. *Ecology*, **35**, 23–30.

Cliff, A. D. and Ord, J. K. (1981). *Spatial Processes: Models and Applications.* London: Pion.

Cook-Mozaffari, P., Darby, S., Doll, R., Forman, D., Hermon, C. and Pike, M.C. (1989). Geographical variation in mortality from leukaemia and other cancers in England and Wales in relation to proximity to nuclear installations, 1969–78. *British Journal of Cancer*, **59**, 476–85.

Cormack, R.M. (1977). The invariance of Cox and Lewis' statistic for the analysis of spatial patterns. *Biometrika* **64**, 143–4.

Cormack, R.M. (1979). Spatial aspects of competition between individuals. In *Spatial and Temporal Analysis in Ecology*, eds. R.M. Cormack and J. K. Ord, 152–211. Fairland: International Co-operative Publishing House.

Cottam, G. and Curtis, J. T. (1949). A method for making rapid surveys of woodlands, by means of pairs of randomly selected trees. *Ecology* **30**, 101–4.

Cox, D.R. (1955). Some statistical methods related with series of events (with discussion). *Journal of the Royal Statistical Society.* B **17**, 129–64.

Cox, D.R. (1972a). The statistical analysis of dependencies in point processes. In *Stochastic Point Processes*, ed. P. A. W. Lewis, 55–66. New York: Wiley.

Cox, D.R. (1972b). Regression models and life tables (with Discussion). *Journal of the Royal Statistical Society* B, **34**, 187–220.

Cox, D.R. (1977). The role of significance tests. *Scandinavian Journal of Statistics*, **4**, 49–70.

Cox, D.R. and Isham, V. (1980). *Point Processes*. London: Chapman and Hall.

Cox, D.R. and Lewis, P. A. W. (1966). *The Statistical Analysis of Series of Events*. London: Methuen.

Cox, D.R. and Lewis, P. A. W. (1972). Multivariate point processes. *Proceedings of the 6th Berkeley Symposium in Mathematical Statistics and Probability*, **3**, 401-448.

Cox, T. F. (1976). The robust estimation of the density of a forest stand using a new conditioned distance method. *Biometrika* **63**, 493–500.

Cox, T. F. (1979). A method for mapping the dense and sparse-regions of a forest stand. *Applied Statistics*, **28**, 14–9.

Cox, T. F. and Lewis, T. (1976). A conditioned distance ratio method for analysing spatial patterns. *Biometrika* **63**, 483–92.

Cressie, N.A.C. (1991). *Statistics for Spatial Data*. New York: Wiley.

Cressie, N. and Huang, H-C. (1999). Classes of non-separable spatio-temporal stationary covariance functions. *Journal of the American Statistical Association*, **94**, 1330–1340.

Cressie, N. and Wikle, C.K. (2011). *Statistics for Spatio-Temporal Data*. Hoboken: Wiley.

Crick, F. H. C. and Lawrence, P. A. (1975). Compartments and polychones in insect development. *Science* **189**, 340–7.

Cuzick, J. and Edwards, R. (1990). Spatial clustering for inhomogeneous populations (with Discussion). *Journal of the Royal Statistical Society*, B **52**, 73–104.

Daley, D. J. and Vere-Jones, D. (1972). A summary of the theory of point processes. In *Stochastic Point Processes*, ed. P.A.W. Lewis, 299–383. New York: Wiley.

Daley, D. J. and Vere-Jones, D. (2002). *Introduction to the Theory of Point Processes: Elementary Theory and Methods*. New York : Springer.

Daley, D. J. and Vere-Jones, D. (2005). *Introduction to the Theory of Point Processes: General Theory and Structure*. New York : Springer.

Diggle, P. J. (1975). Robust density estimation using distance methods. *Biometrika*, **62**, 39–48.

Diggle, P. J. (1977a). A note on robust density estimation for spatial point patterns. *Biometrika*, **64**, 91–5.

Diggle, P. J. (1977b). The detection of random heterogeneity in plant populations. *Biometrics*, **33**, 390–4.

Diggle, P. J. (1978). On parameter estimation for spatial point processes. *Journal of the Royal Statistical Society*, B **40**, 178–81.

Diggle, P. J. (1979a). On parameter estimation and goodness-of-fit testing for spatial point patterns. *Biometrics*, **35**, 87–101.

Diggle, P. J. (1979b). Statistical methods for spatial point patterns in ecology. In *Spatial and Temporal Analysis in Ecology*, eds. R.M. Cormack and J. K. Ord, 95–150. Fairland: International Co-operative Publishing House.

Diggle, P. J. (1981a). Some graphical methods in the analysis of spatial point patterns. In *Interpreting Multivariate Data*, ed. V. Barnett, 55–73. Chichester: Wiley.

Diggle, P. J. (1981b). Statistical analysis of spatial point patterns. *New Zealand Statistician*, **16**, 22–41.

Diggle, P.J. (1985a). Displaced amacrine cells in the retina of a rabbit: analysis of a bivariate spatial point pattern. *Journal of Neuroscience Methods*, **18**, 115–25.

Diggle, P.J. (1985b). A kernel method for smoothing point process data. *Applied Statistics*, **34**, 138–47.

Diggle, P.J. (1990). A point process modelling approach to raised incidence of a rare phenomenon in the vicinity of a prespecified point. *Journal of the Royal Statistical Society*, A **153**, 349–62.

Diggle, P.J. (2006). Spatio-temporal point processes, partial likelihood, foot-and-mouth. *Statistical Methods in Medical Research*, **15**, 325–336.

Diggle, P. L, Besag, J. and Gleaves, J. T. (1976). Statistical analysis of spatial point patterns by means of distance methods. *Biometrics*, **32**, 659–67.

Diggle, P.J. and Cox, T.F. (1983). Some distance-based tests of independence for sparsely-sampled multivariate spatial point patterns. *International Statistical Review*, **51**, 11–23.

Diggle, P.J., Eglen, S.J. and Troy, J.B. (2006). Modelling the bivariate spatial distribution of amacrine cells. In *Case Studies in Spatial Point Processes*, ed A Baddeley, P. Gregori, J. Mateu, R. Stoica and D. Stoyan, 215–233. New York: Springer.

Diggle, P. Elliott, P., Morris, S. and Shaddick, G. (1997). Regression modelling of disease risk in relation to point sources. *Journal of the Royal Statistical Society*, A **160**, 491–505.

Diggle, P.J., Fiksel, T., Grabarnik, P., Ogata, Y., Stoyan, D. and Tanemura, M. (1994). On parameter estimation for pairwise interaction point processes. *International Statistical Review*, **62**, 99–117.

Diggle, P.J., Gates, D.J. and Stibbard. A. (1987). A non-parametric estimator for pairwise-interaction point processes. *Biometrika*, **74**, 763–70.

Diggle, P.J, Gomez-Rubio, V., Brown,P.E., Chetwynd, A.G and Gooding, S. (2007). Second-order analysis of inhomogeneous spatial point process data. *Biometrics*, **63**, 550–557.

Diggle, P.J. and Gratton, R.J. (1984) Monte Carlo methods of inference for implicit statistical models (with discussion). *Journal of the Royal Statistical Society* B, **46**, 193–227.

Diggle, P.J., Kaimi, I. and Abellana, R. (2010). Partial likelihood analysis of spatio-temporal point process data. *Biometrics* **66**, 347–354.

Diggle, P.J., Knorr-Held, L., Rowlingson, B., Su, T., Hawtin, P. and Bryant, T. (2003). Towards on-line spatial surveillance. In *Monitoring the Health of Populations: Statistical Methods for Public Health Surveillance.*, eds. R. Brookmeyer and D. Stroup. Oxford : Oxford University Press.

Diggle, P.J., Lange, N. and Benes, F.M. (1991). Analysis of variance for replicated spatial point patterns in clinical neuroanatomy. *Journal of the American Statistical Association*, **86**, 618–25.

Diggle, P. J. and Matérn, B. (1981). On sampling designs for the estimation of point-event nearest neighbour distributions. *Scandinavian Journal of Statistics*, **7**, 80–4.

Diggle, P.J., Mateu, J. and Clough, H.E. (2000). A comparison between parametric and non-parametric approaches to the analysis of replicated spatial point patterns. *Advances in Applied Probability*, **32**, 331–43

Diggle, P.J., Menezes, R. and Su, T-L. (2010). Geostatistical analysis under preferential sampling (with discussion). *Applied Statistics* **59**, 191–232.

Diggle, P. J. and Milne, R.K. (1983a). Negative binomial quadrat counts and point processes. *Scandinavian Journal of Statistics*, **10**, 257–67.

Diggle, P. J. and Milne, R.K. (1983b). Bivariate Cox processes: some models for bivariate spatial point patterns. *Journal of the Royal Statistical Society*. B **45**, 11–21.

Diggle, P.J., Moraga, P., Rowlingson, B. and Taylor, B. (2013). Spatial and spatio-temporal log-Gaussian Cox processes: extending the geostatistical paradigm. *Statistical Science* (to appear)

Diggle, P.J., Morris, S.E. and Wakefield, J.C. (2000). Point-source modelling using matched case-control data. *Biostatistics*, **1**, 89–105.

Diggle, P.J. and Ribeiro, P.J. (2007). *Model-based Geostatistics*. New York: Springer.

Diggle, P.J. and Rowlingson, B.S. (1994). A conditional approach to point process modelling of elevated risk. *Journal of the Royal Statistical Society*, A **157**, 433-40.

Diggle, P.J., Zheng, P. and Durr, P. (2005). Non-parametric estimation of spatial segregation in a multivariate point process. *Applied Statistics*, **54**, 645–58.

Donnelly, K. (1978). Simulations to determine the variance and edge-effect of total nearest neighbour distance. In *Simulation Methods in Archaeology*, ed. I. Hodder, 91–5. London: Cambridge University Press.

Douglas, J. B. (1979). *Analysis with Standard Contagious Distributions*. Fairland: International Co-operative Publishing House.

Duczmal, L., Kulldorff, M. and Huang, L. (2006). Evaluation of spatial scan statistics for irregularly shaped clusters. *Journal of Computational and Graphical Statistics*, **15**, 428–442.

Du Rietz, G. E. (1929). The fundamental units of vegetation. *Proceedings of the International Congress of Plant Science, Ithaca*, 1, 623–7.

Eberhardt, L. L. (1967). Some developments in "distance sampling". *Biometrics*, **23**, 207–16.

Efron, B. and Tibshirani, R.J. (1993). *An Introduction to the Bootstrap.* London : Chapman and Hall.

Eglen, S.J., Diggle, P.J. and Troy, J.B. (2005). Homotypic constraints dominate positioning of on- and off-center beta retinal ganglion cells. *Visual Neuroscience*, **22**, 859–871.

Elliott, P., Beresford, J.A., Jolley, D.J., Pattenden, S.H. and Hills, M. (1992). Cancer of the larynx and lung near incinerators of waste solvents and oils in Britain. In *Geographical and Environmental Epidemiology: methods for small-area studies*, eds. P. Elliott, J. Cuzick, D. English and R. Stern, 359–67. Oxford : Oxford University Press.

Elliott, P., Wakefield, J.C., Best, N.G. and Briggs, D.J. (2000). *Spatial Epidemiology: methods and applications.* Oxford : Oxford University Press.

Evans, D. A. (1953). Experimental evidence concerning contagious distributions in ecology. *Biometrika*, **40**, 186–211.

Fanshawe, T.F. and Diggle, P.J. (2011). Bivariate geostatistical modelling: a review and an application to spatial variation in radon concentrations. *Environmental and Ecological Statistics* **19**, 139–160.

Feller, W. (1968). *An Introduction to Probability Theory and Its Applications*, Vol. 1, 3rd edn. New York: Wiley.

Finkenstadt, B., Held, L. and Isham, V. (2007). *Statistical Methods for Spatio-Temporal Systems.* Boca Raton: Chapman and Hall/CRC.

Fisher, R.A. (1925). *Statistical Methods for Research Workers.* Edinburgh : Oliver and Boyd.

Fisher, R.A. (1935). *The Design of Experiments.* Edinburgh : Oliver and Boyd.

Fisher, R.A., Thornton, H. G. and Mackenzie, W. A. (1922). The accuracy of the plating method of estimating the density of bacterial populations, with particular reference to the use of Thornton's agar medium with soil samples. *Annals Applied Botany*, **9**, 325–59.

Gabriel, E. and Diggle, P.J. (2009). Second-order analysis of inhomogeneous spatio-temporal point process data. *Statistica Neerlandica* **63**, 43–51

Gardner, M.J. (1989). Review of reported incidences of childhood cancer rates in the vicinity of nuclear installations in the U.K. *Journal of the Royal Statistical Society*, A **152**, 307–25.

Gates, D.J. and Westcott, M. (1986). Clustering estimates in spatial point distributions with unstable potentials. *Annals of the Institute of Statistical Mathematics*, **38 A**, 55-67.

Gelfand, A., Diggle, P.J., Fuentes, M. And Guttorp, P. (2010). *Handbook of Spatial Statistics.* Boca Raton: CRC Press.

Gerrard, D. J. (1969). Competition quotient: a new measure of the competition affecting individual forest trees. Research Bulletin No. 20, Agricultural Experiment Station, Michigan State University.

Geyer C.J. (1999). Likelihood inference for spatial point processes. In *Stochastic Geometry, Likelihood and Computation*, eds. O.E. Barndorff-Nielsen, W.S. Kendall and M.N.M. van Lieshout, 79-140. London: Chapman and Hall.

Geyer, C.J. and Thompson, E.A. (1992). Constrained Monte Carlo maximum likelihood for dependent data (with discussion). *Journal of the Royal Statistical Society*, B **54**, 657–99.

Ghent, A. W. (1963). Studies of regeneration of forest stands devastated by spruce budworm. *Forest Science*, **9**, 295–310.

Gilks, W.R, Richardson, S. and Spiegelhalter, D.J. (1996). *Markov Chain Monte Carlo in Practice*. London : Chapman and Hall.

Gill, P. E. and Murray, W. (1972). Quasi-Newton methods for unconstrained optimization. *Journal of the Institute of Mathematics and its Applications*, **9**, 91–108.

Gneiting, T. (2002). Nonseparable, stationary covariance functions for space-time data. *Journal of the American Statistical Association*, **97**, 590–600.

Gneiting, T. and Guttorp, P. (2010). Continuous parameter spatio-temporal processes. In *Handbook of Spatial Statistics*, eds. A.E. Gelfand, P.J. Diggle, M. Fuentes, P. Guttorp, 427–436. Boca Raton: Chapman and Hall/CRC Press.

Goodall, D. G. (1965). Plot-less tests of inter-specific association. *Journal of Ecology*, **53**, 197–210.

Green, P. J. and Sibson, R. (1978). Computing Dirichlet tessellations in the plane. *Computer Journal*, **21**, 168–73.

Greenland, S. and Morgenstern, H. (1989). Ecological bias, confounding and effect modification. *International Journal of Epidemiology*, **18**, 269–74.

Greig-Smith, P. (1952). The use of random and contiguous quadrats in the study of the structure of plant communities. *Annals of Botany*, **16**, 293–316.

Greig-Smith, P. (1964). *Quantitative Plant Ecology*, 2nd edn. London: Butterworth.

Greig-Smith, P. (1979). Pattern in vegetation. *Journal of Ecology*, **67**, 755–79.

Groendyke, C., Welch, D. and Hunter, D.R. (2012). A network-based analysis of the 1861 Hagelloch measles data. *Biometrics*, **68**(3), 755-765.

Hanisch, K. H. and Stoyan, D. (1979). Formulas for the second-order analysis of marked point processes. *Math. Operationsforsch. Statist. Ser. Statist.*, **10**, 555–60.

Harkness, R.D. and Isham, V. (1983). A bivariate spatial point pattern of ants' nests. *Applied Statistics*, **32**, 293–303.

Hastie, T.J. and Tibshirani, R.J (1990). *Generalized Additive Models*. London : Chapman and Hall.

Heikkinen, J. and Penttinen, A. (1999). Bayesian smoothing in the estimation of the pair potential function of Gibbs point processes. *Bernoulli*, **5**, 1119–36.

Hernández, A. and Ruiz, X. (2003). Predation on common tern eggs by the yellow-legged gull at the Ebro Delta. *Scientia Marina*, **67**, 95-101.

Hines, W.G.S. and Hines, R.J.0. (1979). The Eberhardt index and the detection of non-randomness of spatial point distributions. *Biometrika*, **66**, 73–80.

Ho, L. P. and Stoyan, D. (2008). Modelling marked point patterns by intensity-marked Cox processes. *Statistics and Probability Letters*, **78**, 11941199.

Hodder, I. and Orton, C. (1976). *Spatial Analysis in Archaeology*. London: Cambridge University Press.

Hogmander, H. and Sarrka, A. (1999). Multitype spatial point patterns with hierarchical interactions. *Biometrics*, **55**, 1051–8.

Holgate, P. (1964). The efficiency of nearest neighbour estimators. *Biometrics*, **20**, 647–9.

Holgate, P. (1965a). Tests of randomness based on distance methods. *Biometrika*, **52**, 345–53.

Holgate, P. (1965b). Some new tests of randomness. *Journal of Ecology*, **53**, 261–6.

Holgate, P. (1965c). The distance from a random point to the nearest point of a closely packed lattice. *Biometrika*, **52**, 261–3.

Holgate, P. (1972). The use of distance methods for the analysis of spatial distributions of points. In *Stochastic Point Processes*, ed. P.A.W. Lewis, 122–53. New York: Wiley.

Hope, A. C. A. (1968). A simplified Monte Carlo significance test procedure. *Journal of the Royal Statistical Society*, B **30**, 582–98.

Hopkins, B. (1954). A new method of determining the type of distribution of plant individuals. *Annals of Botany*, **18**, 213–26.

Hsuan, F. (1979). Generating uniform polygonal random pairs. *Applied Statistics*, **28**, 170–2.

Hughes, A. (1981). Cat retina and the sampling theorem: the relation of transient and sustained brisk-unit cut-off frequency to α and β-mode cell density. *Experimental Brain Research*, **42**, 196–202.

Hutchings, M. J. (1979). Standing crop and pattern in pure stands of *Mercu-*

rialis perennis and *Rubus fruticosus* in mixed deciduous woodland. *Oikos*, **31**, 351–7.

Ilian, J., Penttinen, A., Stoyan, H. and Stoyan, D. (2008). *Statistial Analysis and Modelling of Spatial Point Patterns*. Chichester: Wiley.

Jalava, K., Jones, J.A., Goodchild, T., Clifton-Hadley, R., Mitchell, A., Story, A. and Watson, J.M. (2007). No increase in human cases of *Mycobacterium bovis* disease despite resurgence of infections in cattle in the United Kingdom *Epidemiology and Infection*, **135**, 40–45.

Jarner, M.F., Diggle, P.J. and Chetwynd, A.G. (2002). Estimation of spatial variation in risk using matched case-control data. *Biometrical Journal* (submitted).

Kathirgamatamby, N. (1953). Note on the Poisson index of dispersion. *Biometrika*, **40**,225–8.

Keeling, M.J., Woolhouse, M.E.J., Shaw, D.J., Matthews, L., Chase-Topping, M., Haydon, D.T., Cornell, S.J., Kappey, J., Wilesmith, J. and Grenfell, B.T. (2001). Dynamics of the 2001 UK foot and mouth epidemic: stochastic dispersal in a heterogeneous landscape. *Science*, **294**, 813–7.

Kelly, F. P. and Ripley, B. D. (1976). A note on Strauss' model for clustering. *Biometrika*, **63**, 357–60.

Kelsall, J.E. and Diggle, P.J. (1995a). Kernel estimation of relative risk. *Bernoulli*, **1**, 3–16.

Kelsall, J.E. and Diggle, P.J. (1995b). Nonparametric estimation of spatial variation in relative risk. *Statistics in Medicine*, **14**, 2335–42.

Kelsall, J.E. and Diggle, P.J. (1998). Spatial variation in risk: a nonparametric binary regression approach. *Applied Statistics* **47**, 559–73.

Kendall, M. G. (1970). *Rank Correlation Methods*, 4th edn. London: Griffin.

Kennedy, W. J. and Gentle, J. E. (1980). *Statistical Computing*. New York: Marcel Dekker.

Kershaw, K. A. (1957). The use of cover and frequency in the detection of pattern in plant communities. *Ecology*, **38**, 291–9.

Kershaw, K. A. (1973). *Quantitative and Dynamic Plant Ecology*, 2nd edn. London: Arnold.

Kingman, J. F. C. (1977). Remarks on the spatial distribution of a reproducing population. *Journal of Applied Proability*, **14**, 577–83.

Knox, G. (1963). Detection of low intensity epidemicity. *British Journal of Preventive and Social Medicine*, **17**, 121–7.

Knox, G. (1964). Epidemiology of childhood leukaemia in Northumberland and Durham. *British Journal of Preventive and Social Medicine*, **18**, 17–24.

Kulldorff, M. (1997). A spatial scan statistic. *Communications in Statistics Theory and Methods*, **26**, 1481–96.

Kulldorff, M. (1999). An isotonic spatial scan statistic for geographical disease surveillance. *Journal of the National Institute of Public Health*, **48**, 94–101.

Kulldorff1, M., Huang, L., Pickle, L. and Duczmal, L. (2006). An elliptic spatial scan statistic *Statistics in Medicine* **25**, 3929-3943.

Kulldorff, M., Farzad Mostashari,F., Duczmal, L., Yih, W.K.1, Kleinman, K. and Platt, R. (2007). Multivariate scan statistics for disease surveillance. *Statistics in Medicine* **26**, 1824-1833.

Lawson, A.B (1989). Discussion on cancer near nuclear installations. *Journal of the Royal Statistical Society*, A **152**, 374–5.

Lawson, A.B. (1993). On the analysis of mortality events associated with a prespecified fixed point. *Journal of the Royal Statistical Society*, A **156**, 363–77.

Lawson, A. and Leimich, P. (2000). Approaches to the space-time modelling of infectious disease behaviour. *IMA Journal of Mathematics Applied in Medicine and Biology* **17**, 1–13.

Lewis, P. A. W. and Shedler, G. S. (1979). Simulation of non-homogenous Poisson processes by thinning. *Naval Research Logistics Quarterly*, **26**, 403-13.

Lloyd, M. (1967). Mean crowding. *Journal of Animal Ecology*, **36**, 1–30.

Lotwick H. W. (1981). PhD thesis, University of Bath.

Lotwick, H. W. and Silverman, B. W. (1982). Methods for analysing spatial processes of several types of points. *Journal of the Royal Statistical Society*, B **44**, 406–13.

McCullagh, P. and Nelder, J.A. (1989). *Generalized Linear Models* 2nd edn. London : Chapman and Hall.

Ma, C. (2003). Families of spatio-temporal covariance models. *Journal of Statistical Planning and Inference*, **116**, 489–501.

Ma, C. (2008). Recent developments in the construction of spatio-temporal covariance models. *Stochastic Environmental Research and Risk Assessment*, **22**, S39–S47.

Mannion, D. (1964). Random space-filling in one dimension. *Publications of the Mathematical Institute of the Hungarian Academy of Science*, **9**, 143–53.

Marriott, F. H. C. (1979). Monte Carlo tests: how many simulations? *Applied Statistics*, **28**, 75–7.

Matérn, B. (1960). Spatial Variation. Meddelanden fran statens skogs-forsningsinstitut, Vol. 49 (5). Stockholm: Statens Skogsforsningsinstitut.

Matérn, B. (1971). Doubly stochastic Poisson processes in the plane. In *Statistical Ecology*, Vol. 1, eds. G.P. Patil, E.C. Pielou and W.E. Waters, 195–213. University Park: Pennsylvania State University Press.

Matérn, B. (1986) *Spatial Variation.* New York: Springer.

Matheron, G. (1975). *Random Sets and Integral Geometry.* New York: Wiley.

Maynard-Smith, J. (1974). *Models in Ecology.* London: Cambridge University Press.

Mead, R.(1974). A test for spatial pattern at several scales using data from a grid of contiguous quadrats. *Biometrics*, **30**, 295–307.

Møller, J. and Sorensen, M. (1994). Statistical analysis of a spatial birth-and-death process model with a view to modelling linear dune fields. *Scandinavian Journal of Statistics*, **21**, 1–19.

Møller, J., Syversveen, A.R. and Waagepetersen, R.P. (1998). Log-Gaussian Cox processes. *Scandinavian Journal of Statistics*, **25**, 451–82.

Møller, J. and Waagepetersen, R.P. (2004). *Statistical Inference and Simulation for Spatial Point Processes.* London : Chapman and Hall.

Moore, P. G. (1954). Spacing in plant populations. *Ecology* **35**, 222–7.

Morisita, M. (1959). Measuring the dispersion of individuals and analysis of the distributional patterns. *Memoirs of the Faculty of Science, Kyushu University Series E* (Biology), **2**, 215–35.

NAG (1977). *Fortran Library Manual.* Oxford: NAG Executive.

Nelder, J. A. and Mead, R. (1965). A simplex method for function minimization. *Computer Journal*, **7**, 308–13.

Neyman, J. (1939). On a new class of contagious distributions, applicable in entomology and bacteriology. *Annals of Mathematical Statistics*, **10**, 35–57.

Neyman, J. and Scott, E. L. (1958). Statistical approach to problems of cosmology (with discussion). *Journal of the Royal Statistical Society*, B **20**, 1–43.

Numata, M. (1961). Forest vegetation in the vicinity of Choshi. Coastal flora and vegetation at Choshi, Chiba Prefecture IV. *Bulletin of Choshi Marine Laboratory, Chiba University*, **3**, 28–48 [in Japanese].

Ogata, Y. and Tanemura, M. (1981). Estimation of interaction potentials of spatial point patterns through the maximum likelihood procedure. *Annals of the Institute of Statistical Mathematics*, B **33**, 315–38.

Ogata, Y. and Tanemura, M. (1984). Likelihood analysis of spatial point patterns. *Journal of the Royal Statistical Society*, B **46**, 496–518.

Ogata, Y. and Tanemura, M. (1986). Likelihood estimation of interaction potentials and external fields of inhomogeneous spatial point patterns. In *Pacific Statistical Congress*, eds. I.S.Francis, B.J.F.Manly and F.C.Lam, 150–4. Amsterdam, Elsevier.

Ohser, J. and Stoyan, D. (1981). On the second-order and orientation analysis of planar stationary point processes. *Biometrical Journal*, **23**, 523–33.

Ord, J. K. (1978). How many trees in a forest? *Mathematical Scientist*, **3**, 23–33.

Ord, J. K. (1979). Time-series and spatial patterns in ecology. In *Spatial and Temporal Analysis in Ecology*, eds. R.M. Cormack and J.K. Ord, 1–94. Fairland: International Co-operative Publishing House.

Patil, S. A., Burnham, K. P. and Kovner, J. L. (1979). Non-parametric estimation of plant density by the distance method. *Biometrics* **35**, 613–22.

Peebles, P. J. E. (1974). The nature of the distribution of galaxies.*Astronomy and Astrophysics*, **32**,197–202.

Penttinen, A. (1984). Modelling Interaction in Spatial Point Patterns: parameter estimation by the maximum likelihood method. *Jyvaskyla Studies in Computer Science, Economics and Statistics*, **7**. University of Jyvaskyla.

Perry, J. N. and Mead, R. (1979). On the power of the index of dispersion test to detect spatial pattern. *Biometrics* **35**, 613–22.

Persson, 0. (1964). Distance methods. *Studia Forestalia Suecica* **15**, 1–68.

Persson, 0. (1971). The robustness of estimating density by distance measurements. In *Statistical Ecology*, Vol. 2, eds. G.P. Patil, E.C. Pielou and W.E. Waters, 175–90. University Park: Pennsylvania State University Press.

Pielou, E. C. (1977). *Mathematical Ecology*. New York: Wiley.

Pollard, J. H. (1971). On distance estimators of density in randomly distributed forests. *Biometrics*, **27**, 991–1002.

Preston, C. J. (1977). Spatial birth-and-death processes. *Bulletin of the International Statistical Institute*, **46** (2), 371–91.

Prince, M., Chetwynd, A., Diggle, P., Jarner, M., Metcalf, J. and James, O. (2001). The geographical distribution of primary biliary cirrhosis in a well-defined cohort. *Hepatology* **34**, 1083–8.

Rathbun, S.L. (1996). Estimation of Poisson intensity using partially observed concomitant variables. *Biometrics*, **52**, 226–42.

Ripley, B. D. (1976). The second-order analysis of stationary point processes. *Journal of Applied Proability*, **13**, 255–66.

Ripley, B. D. (1977). Modelling spatial patterns (with discussion). *Journal of the Royal Statistical Society*, B **39**, 172–212.

Ripley, B. D. (1978). Spectral analysis and the analysis of pattern in plant communities. *Journal of Ecology*, **66**, 965–81.

Ripley, B. D. (1979a). Tests of "randomness" for spatial point patterns. *Journal of the Royal Statistical Society*, B **41**, 368–74.

Ripley, B. D. (1979b). Simulating spatial patterns: dependent samples from a multivariate density. *Applied Statistics*, **28**, 109–12.

Ripley, B.D. (1981). *Spatial Statistics*. New York: Wiley.

Ripley, B.D. (1987). *Stochastic Simulation*. New York: Wiley.

Ripley, B.D. (1988). *Statistical Inference for Spatial Processes*. Cambridge : Cambridge University Press.

Ripley, B. D. and Kelly, F. P. (1977). Markov point processes. *Journal of the London Mathematical Society*, **15**, 188–92.

Ripley, B. D. and Silverman, B. W. (1978). Quick tests for spatial regularity. *Biometrika*, **65**, 641–2.

Rogers, C. A. (1964). *Packing and Covering*. London: Cambridge University Press.

Rosenblatt, M. (1956). Remarks on some nonparametric estimates of a density function. *Annals of Mathematical Statistics*, **27**, 832–7.

Rodrigues, A. and Diggle, P.J. (2010). A class of convolution-based models for spatio-temporal processes with non-separable covariance structure. *Scandinavian Journal of Statistics*, **37**, 553–567.

Rodrigues, A., Diggle, P. and Assuncao, R. (2010). Semiparametric approach to point source modelling in epidemiology and criminology. *Applied Statistics*, **59**, 533–542.

Rowlingson, B.S. and Diggle, P.J. (1993). Splancs: Spatial point pattern analysis code in S-plus. *Computers in Geosciences*, **19**, 627–55.

Rue, H. and Held, L. (2005). *Gaussian Markov Random Fields: Theory and Applications*. London: CRC Press.

Rue, H., Martino, S. and Chopin, N. (2009). Approximate Bayesian inference for latent Gaussian models using integrated nested Laplace approximations (with discussion). *Journal of the Royal Statistical Society* B, **71**, 319-392.

Sarkka, A. (1983). Pseudo-likelihood approach for pair potential estimation of Gibbs processes. *Jyvaskyla Studies in Computer Science, Economics and Statistics*, **22**. University of Jyvaskyla.

Schabenberger and Gotway, C.A. (2004). *Statistical Methods for Spatial Data Analysis*. Boca Raton: Chapman and Hall/CRC.

Selvin, H.C. (1958). Durkheim's 'suicide' and problems of empirical research. *American Journal of Sociology*, **63**, 607–19.

Schlather, M.S., Ribeiro, P.J. and Diggle, P.J. (2004). Detecting dependence between marks and locations of marked point processes. *Journal of the Royal Statistical Society*, B **66**, 79–93.

Silverman, B. W. (1981). Density estimation for univariate and bivariate data. In *Interpreting Multivariate Data*, ed. V. Barnett, 37–53. Chichester: Wiley.

Silverman, B.W. (1986). *Density estimation for statistics and data analysis*. London: Chapman and Hall.

Silverman, B. W. and Brown, T. (1978). Short distances, flat triangles and Poisson limits. *Journal of Applied Proabability*, **15**, 815–25.

Skellam, J. G. (1958). On the derivation and applicability of Neyman's Type A distribution. *Biometrika*, **45**, 32–6.

Sprent, P. (1981). *Quick Statistics*. Harmondsworth: Penguin.

Stein, M.L. (1991). A new class of estimators for the reduced second moment measure of point processes. *Biometrika*, **78**, 281–6.

Stiteler, W. M. and Patil, G. P. (1971). Variance to mean ratio and Morisita's index as measures of spatial pattern in ecological populations. In *Statistical Ecology*, Vol. 1, eds. G.P. Patil, E.C. Pielou and W.E. Waters, 423–59. University Park: Pennsylvania State University Press.

Stoyan, D. (1979). Interrupted point processes. *Biometrical Journal*, **21**, 607–10.

Stoyan, D., Kendall, W.S. and Mecke, J. (1995). *Stochastic Geometry and its Applications*, 2nd edn. New York : Wiley.

Stoyan, D. and Stoyan, H. (1994). *Fractals, Random Shapes and Point Fields*. New York : Wiley.

Strauss, D. J. (1975). A model for clustering. *Biometrika*, **62**, 467–75.

Tanemura, M. (1979). On random complete packing by discs. *Annals of the Institute of Statistical Mathematics*, **31**, 351–65.

Tango, T. (1995). A class of tests for detecting 'general' and 'focused' clustering of rare diseases. *Statistics in Medicine*, **14**, 2323–34.

Tango, T. (2000). A test for spatial disease clustering adjusted for multiple testing. *Statistics in Medicine*, **19**, 191–204.

Taylor, B.M., Davies, T.M., Rowlingson, B.S. and Diggle, P.J. (2013). lgcp: Inference with spatial and spatio-temporal log-Gaussian Cox processes in R. *Journal of Statistical Software* **52**(4) www.jstatsoft.org/v52/i04

Thomas, M. (1949). A generalization of Poisson's binomial limit for use in ecology. *Biometrika* **36**, 18–25.

Thompson, H.R. (1956). Distribution of distance to nth nearest neighbour in a population of randomly distributed individuals. *Ecology*, **37**, 391–4.

Tildesley, M.J., Bessell, P.R., Keeling, M.J. and Woolhouse, M.E.J. (2009). The role of pre-emptive culling in the control of foot-and-mouth disease. *Proceedings of the Royal Society* B **276**, 3239–3248.

Townsend, P., Phillimore, P. and Beattie, Q.A. (1988). *Health and Deprivation: Inequalities and the North.* London : Croom Helm.

Upton, G.J.G. and Fingleton, B. (1985). *Spatial Data Analysis by Example, Volume 1: Point Pattern and Quantitative Data.* Chichester: Wiley.

Upton, G.J.G. and Fingleton, B. (1989). *Spatial Data Analysis by Example, Volume 2: Categorical and Directional Data.* Chichester: Wiley.

Van Lieshout, M.N.M. (2000). *Markov Point Processes and their Applications.* London : Imperial College Press.

Van Lieshout, M.N.M. and Baddeley, A.J. (1996). A nonparametric measure of spatial interaction in point processes. *Statistical Neerlandica*, **50**, 344–61.

Venables, W.N. and Ripley, B.D. (1994). *Modern Applied Statistics with S-Plus.* New York: Springer.

Waller, L. and Gotway, C.A. (2004). *Applied Spatial Statistics for Public Health Data.* New York: Wiley.

Warren, W. G. (1971). The centre satellite concept as a basis for ecological sampling. In *Statistical Ecology*, Vol. 2, eds. G.P. Patil, E.C. Pielou and W.E. Waters, 87–116. University Park: Pennsylvania State University Press.

Warren, W. G. and Batcheler, C. L. (1979). The density of spatial patterns: robust estimation through distance methods. In *Spatial and Temporal Analysis in Ecology*, eds. R.M. Cormack and J. K. Ord, 240–70. Fairland: International Co-operative Publishing House.

Wichmann, B.A. and Hill, I.D. (1982). Algorithm AS183. An efficient and portable pseudo-random number generator. *Applied Statistics*, **31**, 188–90. (Correction, **33**, 123)

Williams, J.R., Alexander, F.E., Cartwright, R.A. and McNally, R.J.Q. (2001). Methods for eliciting aetiological clues from geographically clustered cases of disease, with application to leukaemia-lymphoma data. *Journal of the Royal Statistical Society*, A **164**, 49–60.

Wood, S.N. (2006). *Generalized Additive Models: An Introduction with R.* Boca Raton: Chapman and Hall.

Woodward, M. (1999). *Epidemiology: Study Design and Data Analysis.* London : Chapman and Hall.

Zhuang, J., Ogata, Y. and Vere-Jones, D. (2002). Stochastic Declustering of space-time earthquake occurrences. *Journal of the American Statistical Association,* **97**, 369–380.

Index